Pine cones close
their scales in
high humidity.

Crickets chirp
faster as the
temperature rises.

Honeybees don't
leave their hives below
55 degrees Fahrenheit.

Cows don't lie down
before rain, but they do head home.

Acclaim for Tristan Gooley

"Gooley interprets clues like a private investigator of the wilds, leaving no stone unturned. . . . For those inclined to solve mysteries written into the landscape, this author's lead is one they'll want to follow." **—The Wall Street Journal**

"[Gooley] has become the global expert on natural navigation, finding his way around the world using nothing but natural clues and pointers. His discovery (made on a sailing expedition to Iceland)—that if, when at sea, you see more than 10 birds in any given five-minute window, this means you are within 40 miles of land—has become part of the British military's survival guidance." **—The Daily Beast**

"Gooley is your man. . . . With unflappable practicality, he shares simple ways to understand your surroundings, whether you're beside a stream or on the open sea at night, without instruments." **—Discover magazine**

"While Gooley's tips encompass useful, practical ways to predict a change in weather, determine when a predator may be prowling and find true North at night, his true gift is in igniting curiosity and wonder about the world around us." **—Shelf Awareness**

"Avid and budding outdoorspeople will appreciate Gooley's breadth of knowledge and accessible approach." **—Publishers Weekly**

"Gooley, who has single-handedly been reviving natural navigation in this age of GPS, has the birdwatching skills of Bill Oddie and the deductive powers of Sherlock Holmes. He can make you feel that you've spent half your life walking about with your eyes only half-open." **—The Telegraph**

The Secret World
of Weather

The Secret World of Weather

of Weather

How to Read Signs in Every Cloud,

Breeze, Hill, Street, Plant,

Animal, and Dewdrop

TRISTAN GOOLEY

Illustrations by Neil Gower

THE EXPERIMENT

NEW YORK

THE SECRET WORLD OF WEATHER: *How to Read Signs in Every Cloud, Breeze, Hill, Street, Plant, Animal, and Dewdrop*
Text and photographs copyright © 2021 by Tristan Gooley
Illustrations copyright © 2021 by Neil Gower

Simultaneously published in the UK by Sceptre. First published in North America in revised form by The Experiment, LLC.

The Experiment, LLC
220 East 23rd Street, Suite 600
New York, NY 10001-4658
theexperimentpublishing.com

THE EXPERIMENT and its colophon are registered trademarks of The Experiment, LLC. Many of the designations used by manufacturers and sellers to distinguish their products are claimed as trademarks. Where those designations appear in this book and The Experiment was aware of a trademark claim, the designations have been capitalized.

The Experiment's books are available at special discounts when purchased in bulk for premiums and sales promotions as well as for fundraising or educational use. For details, contact us at info@theexperimentpublishing.com.

Library of Congress Cataloging-in-Publication Data

Names: Gooley, Tristan, author. | Gower, Neil, illustrator.
Title: The secret world of weather : how to read signs in every cloud, breeze, hill, street, plant, animal, and dewdrop / Tristan Gooley ; illustrations by Neil Gower.
Description: New York : The Experiment, 2021. | Includes bibliographical references and index.
Identifiers: LCCN 2021001387 (print) | LCCN 2021001388 (ebook) | ISBN 9781615197545 (hardcover) | ISBN 9781615197552 (ebook)
Subjects: LCSH: Weather forecasting.
Classification: LCC QC995.48 .G66 2021 (print) | LCC QC995.48 (ebook) | DDC 551.63--dc23
LC record available at https://lccn.loc.gov/2021001387
LC ebook record available at https://lccn.loc.gov/2021001388

ISBN 978-1-61519-754-5
Ebook ISBN 978-1-61519-755-2

Cover and endpaper design and illustration by Beth Bugler
Text design by Jack Dunnington
Author photograph by Jim Holden

Manufactured in the United States of America

First printing May 2021
10 9 8 7 6 5 4 3 2 1

For Sophie, on our 30/20 Anniversary.
Thank you for always being there,
whatever the weather.

Contents

To see additional images of this book's concepts as they appear in the wild, please visit: naturalnavigator.com/news/tag/secret-world-of-weather.

Introduction

This is an unorthodox book about the weather.

My investigation sidesteps the charts on screens and instead focuses on the clues we find as we walk around a tree or down a street—and what they reveal to us about our present, past, and future weather. This path will take us deep into a little-explored but wonderful realm: microclimates. It is time to rejoice in small local observations and celebrate the weather signs that few notice. They are out there, in the sky and throughout our landscapes, waiting for us. Many are within touching distance.

I hope you enjoy the journey.

(Note: Unless otherwise stated, we are considering the north temperate zone, which includes most of the populated parts of Europe, North America, and Asia.)

Tristan

CHAPTER 1

Two Worlds

The Known World • The Secret World • The Blocking High •
The Tree Fan

I T WAS A LATE SEPTEMBER DAY, very warm
with light breezes. Summer held on to the land. I walked
past an oak tree I knew well and looked out, below a
bright sun, over the green hills of the South Downs. They
wobbled in the heat. There were a few puffy clouds low in the
sky and none higher up. Visibility was not great, but the sea
was there, a dull dark band in the distance.

It was Thursday, and we wanted to go for a family picnic
on the weekend. I felt the breeze on the nape of my neck,
looked back at the oak and its shadow, and knew then that
the weather would hold. I could see the perfect spot for the
picnic on Sunday.

There are several clues and two signs contained in this short,
plain story. In different ways each can help us to understand
what the weather has done and will be doing. But more important, they show us the way into the secret world of weather.

THE KNOWN WORLD

Weather forecasts have gained a bad name, and it started early. Robert FitzRoy, a nineteenth-century vice admiral in the Royal Navy, was a weather pioneer and the person who coined the word "forecast." His reward for attempting to forge new methods in a difficult area? The public bombarded him with criticism following any incorrect forecast. It was hard to bear. FitzRoy grew depressed, and took his life in 1865.

He was ahead of his time. In the same year that he committed suicide, the learned bods at the Royal Society offered their thoughts on weather forecasting: "We can find no evidence that any competent meteorologist believes the science to be at present in such a state as to enable an observer to indicate day by day the weather to be experienced in the next forty-eight hours."

One hundred years later, by the mid-twentieth century, weather forecasts were routine, but doubts lingered. The chief forecaster at Central Forecasting Station in Dunstable, Bedfordshire, sounded less than confident in 1955: "Very little degree of accuracy can be guaranteed for any forecast issued more than twenty-four hours ahead."

And yet, seventy years later, it takes only seconds to find several forecasts that claim to know what our weather will be doing in ten days' time. How so? Did we grow better at learning to read the signs in the sky? In a word, no.

Over the past century, there has been a revolution in four areas: We have much more plentiful and accurate data, a better understanding of the processes that govern weather, formidable data-crunching machines, and swift communication. Readings from across the globe and at every level, from high in the atmosphere to the temperature of the sea, are plugged into computers that spit out their prognostications.

Communication is more important than we might guess. It doesn't do any good to measure the air pressure in the middle of the Atlantic if it leads to a forecast that takes a fortnight to reach someone on this side of the ocean. It's hard to believe that less than a century ago many people in coastal areas relied on cones being hoisted up a mast to warn them that gales were on their way. Even if somebody had cracked a way to forecast the weather accurately a few days ahead, it would take too much cone-hoisting to send the message across a territory.

There are moments when we can look back to see the shift happening, sometimes all too slowly for the poor souls who witnessed the weather. Shortly before the Second World War, a gale rose over a previously calm sea off the west coast of Ireland. Forty-four fishermen died soon afterward. Many miles away forecasters had predicted the storm and issued warnings by radio, but they did not reach as far as the islands off County Mayo.

I have mentioned ten-day forecasts. There is a big difference between making forecasts and making dependable ones. Experience leads me to believe that even the mighty supercomputers struggle beyond five days: Their bold predictions become markedly less reliable by days six and seven. Now, though, we are at the stage when a five-day forecast has value. Two decades ago I gave little time to forecasts that ranged beyond three days. Things are improving quickly and in many areas, but not all.

The developments in professional forecasting have led to a strange relationship between ourselves and the weather. First, most people have lost the belief that we can look at the weather as the source of its own forecast. Second, weather has become detached from its home: the land.

There is now an imbalance between how professionals describe weather and how we experience it. You will have noticed that TV and internet forecasts contain vast swirls that cover entire regions.

It might take five hours to drive across a single forecast region, yet we experience weather on a much smaller scale.

If a meteorologist speaks of "showers" in conversation, I like to ask whether it will rain in my back yard. This often prompts a laugh, because they know all too well where I'm going: They know the limits of their approach. If the hundred best meteorologists in the world borrowed a hundred of the world's most powerful computers, they would still struggle to work out exactly where a predicted shower will fall tomorrow. And they will concede total defeat if they don't know the landscape intimately. These are wise people and they are doing amazing things, but when it comes to the scale in which we actually experience the weather, they are up against it. A forty-eight-hour forecast was deemed impossible in 1865, and forecasting accurately on a small scale remains impossible for computers that don't know the land.

The same need not be true for those of us who rely on our senses. We may struggle to predict shifts in weather trends five days ahead, but we can often tell exactly where rain will fall later in the day. We have an unfair advantage over meteorologists in this game for two reasons. First, they are catering to thousands over a wide area, while we are more interested in how the weather affects us than how it affects anyone in a neighboring county. Second, they treat the weather mainly as an atmospheric phenomenon, but we experience it as creatures of the land it envelops.

A person sensitive to their landscape is granted powers of understanding denied to machines.

THE SECRET WORLD

Our landscape shapes our weather.

The computers are quite happy to factor in large land masses, but they don't trouble themselves by asking how the weather will vary as we walk around a small local hill. The sun, wind, rain, temperature, and visibility can fluctuate significantly on any short walk. This is what we have always meant by "weather," and it is different on two sides of a tree. This is a basic truth, yet if you suggest it to a professional meteorologist they will demur: "Ah, that's not really weather. What you're talking about there is *microclimate*."

I have heard a version of this reply many times, and I always agree, saying, "Yes." But that answer conceals this thought: Give it any name you like. I'm talking about the weather we actually experience.

We live in cities, on hills, in valleys, by the coast, in woods, on islands. We live in a landscape that is shaped by the weather and, in turn, shapes the weather. Woodlands lead to more rain, which helps many tree species to live in that space, and the cycle is strengthened. Woods are a basic sign that rain is more likely there than in the nearby area without trees. And the rain we feel changes as we walk from one tree species to the next.

A small, flat island has different weather from a neighboring large, hilly one. And that larger island experiences different weather on each side. Viewed from above, many islands are completely different colors on either side: One side receives nearly all of the rain and the other almost none. On the same day we might find sunbathers sizzling on the dry southwest coasts of the Canary Islands but rain-soaked plants on the opposite northeast.

The more we zoom in on any landscape, the more striking the shifts we find. The climate on two sides of a 2,600-foot-high (800 m) ridge in the Swiss Jura mountains is so different that two separate ecosystems almost touch each other. Trees that need warm conditions, like downy oak, are found on the south slope, and subalpine species, like Alpine pennycress, on the north. The two environments are separated by a ridge that is less than 2 feet

(50 cm) wide. In climate terms, we can walk across a change similar to about 625 miles (1,000 km) in latitude or more than 3,000 feet (1,000 m) in altitude *in a single step*. And that, by definition, means that the weather is, on average, wildly different over such small distances, too, and predictably so.

The difference in climate between the north and south sides of juniper bushes in temperate zones of the US and Europe is as stark as that between desert and a northern forest. Scientists found that the microclimate around these bushes varied by the same amount over a few yards as the broader climate did over 3,000 miles (5,000 km). When exploring these bushes our arms can stretch across a continent of weather.

I must emphasize that these are not theoretical differences, not just academic facts or measurements. Microclimates reveal average and probable weather conditions, but they also dictate them. They give us clues to what we will experience. Once we recognize how habitats reflect and change the weather, it is exciting to predict and then feel those changes.

On a walk in early December, I crossed heathland under the stars. As I stepped out from under some pines, I expected and felt a sudden chill in the air. Then I spotted frozen puddles among the heather, but none in grasslands or in the woodlands nearby. It was joyous to sense these things and especially satisfying to understand why. Heathland loses heat very quickly at night and can easily be around 5°F (3°C) colder than habitats only a few hundred yards away. (In the next chapter, we will learn why heathland loses heat so quickly.)

Meteorologists know about these wild differences on small scales and loathe them, so much so that they always try to position their anemometers and thermometers at a height that spares them these fluctuations. However much scientific sense this makes, it is ironic that weather forecasters like to measure variables, such as wind and temperature, above the level at which we experience them.

Forecasters have developed an amazing understanding of the weather on the large scale: they have given us a "known world" of big weather. They have done great work that has saved countless lives. But it has had some unintended consequences: Their success has led us to think of the weather on a scale far larger than the one we inhabit.

In this book we will explore the clues and signs that unlock the weather we experience in the towns and cities and out among the trees and hills. Some of these signs point to large-scale events and overlap with the known world of meteorologists, but most are nestling in the landscapes we inhabit. And quite a few are within touching distance. This is the secret world of weather.

Let's start by looking at the clues and signs from my walk earlier: They will help us to celebrate this difference and point the way from the known into the secret world.

THE BLOCKING HIGH

On my walk at the start of this chapter, there was a sunny sky with a few low clouds, no high ones, and light breezes. The warm air wobbled, and the visibility was okay but not great, the details muted. These are all clues, the hallmarks of a summer high-pressure system.

When an area of high pressure sits over a region in summer, it heralds sunny, settled weather; light, variable winds; less than brilliant visibility; and few clouds. For as long as the system remains over the area, this weather continues. Some high-pressure systems are quite stubborn—they squat in one place and don't easily budge. They're known as "blocking highs" and lie behind most heat waves. So when we've worked out that we're sitting in one, all we need to do is keep an eye on where it is relative to us. This will tell us how long the good weather will hold.

It's easy to track a high-pressure system: We just need to keep tabs on wind direction. Winds circle clockwise around these systems, so if you have the wind on your back the high-pressure system will be to your right. In the case above, I felt the wind on my back as I looked downhill. I knew I was looking south but confirmed it by checking the shadow of that oak tree: It was near the middle of the day, the sun was due south, and the shadow pointed north. So the center of the high pressure was to the west.

The Earth rotates toward the east, which means most winds blow in that direction, too, flowing from west to east. This, in turn, means that most weather arrives from the west.

Bringing the pieces together, I could sense I was in a high-pressure system and could tell from the breeze that its center was to the west of me. This was a sign that a large fair-weather system had only just begun its slow march across the land. The fine weather would hold through the weekend and would get better before breaking.

Don't worry about the details yet: We'll meet this blocking-high character again and get to know it better. But for now I'd encourage you to notice one simple thing. Start taking an interest in the wind direction as soon as a period of good, sunny weather settles in. There will be light, sometimes variable breezes during these spells, but note how the wind direction changes *before* these halcyon periods of good weather give way to unsettled skies.

The blocking high is a big bold sign and a system large enough that we can see it on the forecaster's swirls and circles. It is a useful sign and overlaps with the large known world of weather. But now let's step into the secret world and meet the type of sign you won't see in any weather forecast, ever.

THE TREE FAN

The perfect picnic spot I chose at the start of the chapter was under the oak tree. We have all stood under trees to keep cool on very hot days but, bizarrely, very few people know exactly why they do this. It's true that the main reason is to take advantage of the shade, but there's a secret reason, too. We enjoy the breeze there.

When any wind passes a tree, the tree gets in its way. This causes changes in the air pressure on all sides of the tree. The pressure increases on the upwind side and decreases on the downwind. The higher pressure on the windward side then accelerates over, around, and under the tree. This leads to a breeze under a tree that is faster and stronger than the breeze away from the tree. We feel cooler under trees on hot days because of the shade *and* the cooling breeze.

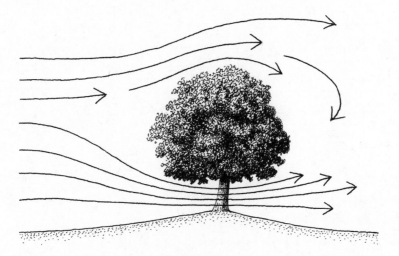

The Tree Fan

The two signs we have just met mark out our territory. They are near each end of the spectrum of signs we will be exploring. The blocking high is a sign of a massive weather system; it is from the known world and gives an understanding of what the weather will be doing over several days and hundreds of miles. The tree fan is from the secret world. It's a microclimate sign, hyperlocal, but also immediate and reliable. Together they paint a useful and fascinating picture of the weather we will experience.

The weather held for long enough. We enjoyed our weekend picnic in the breeze under the oak, near bleating sheep and cawing crows. The wind direction changed on Sunday, and on Monday it clouded over and cooled down.

CHAPTER 2

The Secret Laws

INTERPRETING WEATHER SIGNS with confidence requires knowledge of what is going on out there. We need to look at the building blocks behind the signs, and we'll do that by using our senses. This chapter is about the hidden logic behind what we see. You may find parts of it challenging at first. Don't feel you need to nail these concepts instantly; we'll meet them many times and get to know them well. And once you start to see them at play in nature, they grow much friendlier. I promise you'll soon be able to spot and decipher beautiful weather patterns and continue to do so for the rest of your life, thanks to the ideas in this chapter.

THE ART OF SITTING IN A SUN POCKET

Weather is a soup of heat, air, and water, stirred constantly by the sun and the rotation of our planet. It never heats uniformly or mixes perfectly, which is why no two days are identical. Every weather phenomenon can be broken down into those three ingredients: heat, air, and water. The better we know the ingredients and how they behave, the better we get at reading the weather. Let's start with heat.

How pleasurable it is to feel warmed by the sun on a winter's day! On a still, cold day, when ice and a sprinkling of snow coat the land, who has not enjoyed feeling the sun warming their face? Such is the joy of its radiating power.

We know that the sun's radiating energy reaches our planet and warms it. And we also know that the more energy that reaches a region, the warmer it is likely to be. We expect to swelter on a summer afternoon or when we're near the equator, just as we anticipate shivering in winter, on a cold night, or at high latitudes.

After feeling the sun on your face on a cold morning, you look around and see that it has thawed some parts of the frosty ground, but not others. You touch the different parts of your jacket, and the dark parts feel much warmer than the light—some are even hot. Yet the air is freezing and you can see your breath.

Energy has radiated from the sun, passing through the air to your face and your clothes and heated them, but not equally. It has warmed your face a little, the dark part of your jacket more, the light part less, and the air less again. You will be familiar with the uneven absorption of the sun's energy from hot days, too, when a dark car's hood feels much hotter to the touch than that of the white one parked next to it. Dark colors absorb more of the sun's radiation than light ones.

On a cold, frosty day we can still find a warm, comfortable place to sit. We just need to seek out the spot that receives the most direct radiation, maximizes absorption, and doesn't let the precious heat escape. This is the art of finding a good sun pocket. The south side of woodland on a slope is a great place, especially if there are overhanging branches that let the low sun all the way in but shield out the sky directly above you.

The slope tilts you toward the sun, and the tree cover helps in three ways. The woods will shelter you from the wind, but the overhanging branches also protect the ground from snow and frost, leaving it darker to absorb more of the sun's radiation. The branches stop some of this heat escaping upward, trapping it in the pocket. It can feel like an odd thing to do, to head under tree branches on a sunny day when you're cold, but the temperature difference between a pocket and standing out in the sun in the open is considerable. If you need to keep still outdoors for long periods in cold weather, sun pockets are worth seeking out. The animals know them well.

You can, of course, achieve exactly the same effect by sitting under the eaves of a jutting roof. In the Alps it's common to spot people reading under one, quite warm and comfortable, as steam from the breath of those walking past curls gently upward in the freezing air.

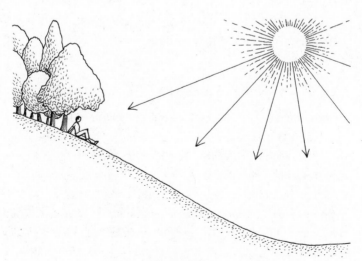

The Sun Pocket

There are sources of radiating energy all around us, all the time. Everything above −459.67°F (−273°C; absolute zero) emits the same invisible infrared energy that arrives from the sun. And since everything on our planet is much warmer than that, everything around us is radiating energy and heat to its surroundings. The hot drink next to you, the ground by your feet, or the tree a few hundred feet away are all radiating energy that warms you. True, it's only a minuscule amount when compared to a generous serving arriving from that ball burning at more than 9,000°F (5,000°C) millions of miles away, but it's there and it has an impact.

Every habitat has its own heat fingerprint. It absorbs and loses heat through radiation in its own unique way. In the last chapter we felt a chill in cold heathland. This is a landscape that radiates heat away especially quickly, which is why the puddles were frozen there and not nearby.

Radiation can be hard to visualize. It leaps millions of miles across the vacuum of space to heat the planet during the day, then flows out of the cold land back to space at night. We'll come to know it much better when we spend time with some of radiation's colleagues, like frost and dew patterns.

COLD KNIVES AND WARM SPOONS

The second way heat moves from one place to another is the simplest and most familiar. When something warm touches something cold, heat will flow from the hot object toward the cold one. The vibrating molecules bump into those in the neighboring material and pass on their agitation. This is conduction.

Some materials conduct heat better than others. Metal is a good conductor, and wood is bad. That is why a metal knife feels colder than a wooden spoon, even if they are in the same kitchen drawer at exactly the same temperature. The metal in the knife allows the heat to flow quickly from your fingers, giving you a cool sensation; the wood does not.

Water is a better conductor of heat than air is. Walking in air that feels cool but comfortable can quickly become a survival situation if you fall into water because your body heat will flow much more quickly into it. Sleeping bags keep you warm on cold ground by being really bad conductors of heat and blocking that heat transfer to the ground or air.

If you ever need to spend a night on bare earth it's worth experimenting. Rocks conduct heat better than sand, which conducts it better than peat. But every soil has its own thermal character, and the water content in the ground will have a big impact.

Wild pigs are wise to this game. They've worked out that anthills are very bad conductors of heat, so they break them down to use as a thermal blanket. And that is where we can leave conduction: It has a greater impact on our experience of the weather than on the weather itself.

SEEDS, SILK, AND WINGS

Warm air expands, making it less dense than cold air. Warm air rises above cool air. Bonfire smoke and kettle steam rise through cooler air. This is convection: the third way in which heat moves from one place to another.

Did you notice earlier how the Alpine walkers' breath curled upward? Our body temperature averages around 98.6°F (37°C), which is nearly always warmer than the air around us. On cold days, our breath rises above us.

The air is transparent, so most convection is invisible but, like radiation, it is taking place all around us. As soon as the sun rises, its radiation warms the land, and the air immediately above the land heats and begins to rise. When a column of warm air rises through convection it is known as a thermal. The column might be a few yards or a few hundred yards wide.

The first powered aircraft were fragile, and every flight was perilous. The last thing these adrenaline-soaked pilots needed was turbulent air, so they preferred to attempt to fly near the start of the day, before the sun could warm the land and cause thermals. Aircraft design has come a long way, and modern airplanes are rarely threatened by convection, but we still feel it as bumps in the ride during the early and late stages of flights. That gentlest of jolts you feel as you pass through the low cloud layer is convection.

The sun heats the land unevenly and at differing times—east-facing slopes warm long before west-facing ones because the sun rises in the east—and the warmest areas generate the strongest thermals. This is partly about the nature of the landscape, but also about time and angles. A hill can leave a valley in shade until the afternoon, and when the sun does reach in, it will heat the woodland, river, fields, town, and lake differently. Dark, dry areas heat faster than light, moist ones. The sun causes the thermals, but the landscape determines where they rise.

In the eighteenth century, the ever-percipient naturalist Gilbert White observed "small spiders which swarm in the fields in fine weather in autumn, and have a power of shooting out webs from their tails so as to render themselves buoyant and lighter than air." And in the 1830s, Charles Darwin noticed spiders reaching his ship, the *Beagle*, even though it was one hundred yards off Argentina at the time. The spiders are not lighter than air, as White supposed: They are flying, or "ballooning," using their silk as a sail and hitching a lift on thermals. The rising air can carry the spiders across continents, but their wonderful journeys are usually much shorter—you may have seen their silk sails as a glistening patchwork covering grass in autumn. This is much more likely following the type of weather spiders favor: sunshine and very light winds. Interestingly, there is some evidence that if the heating of the land is too intense and the thermals too powerful, the spiders wait until things have calmed down a little before embarking.

Animal-behavior experts have known for at least a century that bird weight, thermals, and time of day are all related. Birds of prey climb on thermals—if you notice a bird gaining height as it keeps circling over the same dark woodland on a sunny day, you are watching convection in action. At sunrise there are no sun-driven thermals, but as the morning progresses the land warms and they start to kick in. The bigger the bird, the stronger the rising air current it needs, so we are more likely to see small, light birds of prey circling before their bigger cousins. I like to think of this as a clock: The birds fly around and around, getting bigger as the day grows older. Circling birds map the rising air currents and the warm land beneath them.

After the first light of day, before the thermals are strong enough for even the smallest birds to climb, you may spot dragon-flies making good use of the weakest currents. And before the dragonflies get going, the seeds take off. Many species of plants produce airborne seeds that depend on thermals for their sur-vival. Gravity is always trying to pull the seeds straight to the ground, but they won't survive if they start life in the shade of their parent. They have no independent means of propulsion or climbing, so for many species convection is the difference be-tween landing too close to home and the start of a successful life on a new patch.

Happily, we can watch many such seeds on their journeys, but it helps to have the right light. Sunlight passing through a small window into a dark room can highlight the smallest specks of dust floating in the air, and we experience a similar effect in nature. A sunbeam cutting a thin, bright slice between two dark trees works well. Instead of looking for the perfect lighting conditions, try to stay sensitive to spotting airborne seeds—at certain times of the year the air is thick with them, as hay fever sufferers will confirm. You will spot the biggest, fluffiest ones, like ragwort, most easily, but the species doesn't matter. If it is airborne and visible, it is trying to map the gentlest of air currents for you.

Keep watching the seeds and you'll spot how they ride sideways on any breeze, which is interesting, and something we'll come back to in chapter 14, The Trees, but for now we want a very calm day. Track your target carefully and you will spot the moment a seed begins a dramatic climb, the closer to vertical the better. It has stepped into a lift made of air, a thermal. Look closely at the spot below where this happens and you will see how and why it has crossed into a thermal. The ground is likely to be better lit, darker, or drier than the patch next to it, which is why it is warmer.

Congratulations! You have mapped previously uncharted territory. By taking the time to observe this seed and tiny thermal, you have doubtless become the first person in history to have noticed how the air currents behave above that small patch of land.

THE STEAMY ROADS OF SATURATION

If we place an ice cube on a table in the sun we know what will happen next. It will change state from solid to liquid. It melts. If we come back a few hours later there may be nothing left on the table. The water has changed state again, this time from liquid to gas. It has evaporated, becoming water vapor. Some of the ice will have skipped the middle step and sublimated, changing from a solid straight into a gas, but there will be few surprises in what we see.

Water's journey in the other direction happens as regularly but is less familiar. It's rare any of us can claim to have been watching as water turned into ice. However, we see gaseous water vapor condense into steam quite regularly—our breath on a cold day, above fresh coffee, or from car exhausts. A lot of interesting but overlooked things happen when water changes state.

The change from gas to liquid—condensation—is most significant to weather readers. All air on our planet holds some water in gas form: There is no perfectly dry air anywhere, even over the driest deserts on Earth. The warmer the air, the more water it can hold as a gas, as vapor. When air can no longer hold any more water vapor it is "saturated." Saturation occurs when either the vapor levels rise or the temperature drops to the point at which air cannot hold any more water as vapor. This is the moment at which the vapor condenses to liquid and when we start to see the water: clouds, fog, and steam form as a result of this process. Remember, if we can see the water, it must be liquid or solid. Water vapor, the gas, is invisible.

The same is true in reverse: If air dries or warms, fog or clouds disappear, returning to transparent gaseous water vapor. Every time we see fog or a cloud appear or disappear, we are watching this process.

The temperature to which air needs to cool to reach saturation is called the "dew point." The more water in the air, the higher the dew point temperature, and vice versa. In other words, very wet air will form clouds if it cools a little, but very dry air needs much colder temperatures.

After rain has passed and the sun comes out on a cold day, look for steam rising off plants or a road. The sun has warmed the rainwater until it has evaporated as vapor, which is invisible. But it has quickly cooled in the air above, reached the dew point and saturation. The vapor has condensed, returning to water and giving us the snaking threads of steam.

In summary, temperature is the key. If it drops, vapor is more likely to turn to visible water. We see steam rise off ponds on cold mornings but no fog on hot days.

APPLES ON A ROLL—STABILITY

Every morning I gaze at the sky and ask myself, *Does it look stable?* It's such an important habit, one that reveals so much in a few seconds. Once you understand how to go about it, it quickly becomes automatic. You will be able to do it by the end of the chapter.

There are many scenarios in nature that alternate between periods of stability and the reverse. If the numbers of prey animals—like rabbits—increase, their predators (the foxes) do well, breed successfully, and eat more rabbits. The rabbit population falls. If the foxes eat too many rabbits, rabbit numbers plunge, the foxes starve, the rabbit population grows, and the cycle starts again. In this scenario, the population of rabbits could be said to be fairly stable: Any changes lead to further changes that bring things back to what they were nearer the starting point. But if disease killed all the foxes, the number of rabbits might shoot out of control for a time and the population would be unstable.

In stable systems, a change creates forces that bring things back toward the starting point. In an unstable system, a change leads to more change, which sets up a loop of greater changes still. Here's an unscientific definition of unstable situations: One small change and it all kicks off.

We can experiment with stability at home. Put a large bowl on the kitchen table and place an apple in the bottom. If you push the apple a little way up the side of the bowl and let go, it rolls down to its starting point. You can nudge the apple up the side a hundred times and the "system" will return to its starting point. It is stable.

Next, turn the bowl upside down and place the apple on top. Now give it the same little push and watch as it rolls fast down the side of the bowl, off the kitchen table, and along the floor. One small nudge and mayhem ensues. This is an unstable system.

The atmosphere is either stable or unstable, and to understand that, we need to grapple with latent heat. For my money, the release of latent heat is the least-known hugely important weather process. It takes a moment to crack, but once we're comfortable with it, we can understand why and how one of the simplest, most powerful weather signs works. We can then almost instantly tell whether the weather around us is stable or unstable, and why. And this is something we need to be able to do. It is well worth the effort.

THE HUMID BLANKET—LATENT HEAT

Energy can move and change form, but it never disappears. The total amount of energy is always conserved. This is a law of the universe, and we don't break those.

Water vapor has more energy than liquid water, which in turn has more energy than ice. It follows that if vapor turns to water, or water turns to ice, this energy has to go somewhere, and it does: It is released to the air as heat. This heat has its own name, "latent heat" (from the Latin *latere*, meaning "to lie hidden"). The same process works in reverse: Whenever ice turns to water, or water turns to steam, latent heat is absorbed by the water.

Liquid water doesn't get colder than 32°F (0°C) or hotter than 212°F (100°C).* The reason that both temperatures are dependable is because, at the transition from one water state to another, all

---

* Both temperatures change if the air pressure or water purity changes, but that's irrelevant here.

of the extra energy goes into the change of state—turning boiling water into gas, or ice into water. None goes into changing the temperature of the water.

Imagine a large bowl filled with water and lots of ice cubes. If you measure the temperature of the water it will be 32°F. If you warm that bowl a little, the ice will melt, and if you put it in the freezer for a few minutes, more ice will form, but in neither case does the temperature of the water itself change. It stays at 32°F. If you heat boiling water more strongly, you get more steam, not hotter water. If you see a cooking recipe with instructions for a "fast boil," don't waste your gas: The water won't go any higher than 212°F, you'll just make the kitchen steamier.

Now we're ready to see what all this has to do with the weather signs we observe. When the air reaches saturation and condenses, a cloud forms. But this means energy must be released as heat. Cloud forming actually heats the air. And as we know, warm air expands and becomes less dense. So when a cloud forms, it releases heat that in turn makes the air warmer, less dense, and more likely to rise. This may lead to more expansion, more cooling, more saturation, and more cloud forming. The energy released by the water vapor as a cloud grows can start the process all over again.

In theory, this process could go on forever, but most of the time the latent heat released when clouds form isn't enough to keep the process going, and things quickly peter out. This gives us clouds that are wider than they are tall. Sometimes the heat released is enough to keep the process going and the cycle starts to feed itself, which creates clouds that are taller than they are wide. If this keeps happening, it can run away with itself, and when it does, we have a special name for it: a thunderstorm.

The amount of heat that is released by a cloud forming is a constant, so why do some clouds grow tall and menacing and others form flatter, kinder characters? The answer lies in the way the atmosphere's temperature changes with altitude. If the air cools faster with height than the clouds cool as they rise and expand, the clouds will keep rising for as long as this remains true. This is an "unstable atmosphere." Any time warm air is beneath cooler air, the atmosphere is unstable. If the air cools gradually with altitude, clouds rise a bit, then stop and flatten out. This is a stable atmosphere.

We can't expect to see stability or instability directly, as the air is invisible, but the clouds act as markers. If they are growing vertically, becoming much taller instead of wider, and not reaching an obvious ceiling, the atmosphere is unstable. The apple is ready to roll away. The stability of the air has a bearing on almost every cloud sign we will be looking at.

We have all felt an unstable atmosphere, on those hot, humid late-summer afternoons when we expect the weather to "break." It never feels like that in dry heat. What we are sensing indirectly is the power of latent heat. Warm, very humid air rises and condenses, forming clouds, but because the air is very humid, so much condensation is happening that a lot of latent heat is released. This adds fuel to the fire: The clouds continue to rise and a thunderstorm is brewing.

Have you ever had that feeling after a warm, muggy day that the night is going to stay uncomfortably warm, too? You're picking up a sign I nickname the "humid blanket."

When the air is very humid, it is unlikely that the night will suddenly grow cold. This is because the latent heat in the water vapor acts as a brake on any cooling. If humid air starts to cool, it quickly reaches the dew point and the vapor starts to condense, releasing plenty of latent heat, which keeps the air temperature from dropping very much. The latent heat in the vapor creates a humid blanket.

THE GLASS CEILING

Have you noticed how clouds sometimes look as if they're trapped under a glass ceiling?

The atmosphere normally gets cooler with height. We're used to this—we expect snow at the tops of mountains, not the bottom. But this trend does not always hold true. Sometimes a layer of warmer air sits on top of cooler air. This acts as a cap on the layers below.

If rising air hits a layer of warmer air, it abruptly stops rising. It has hit a glass ceiling known as a temperature "inversion," and it spreads out under this layer of warmer air. The atmosphere is super-stable at this level.

We will meet this effect regularly—it is common and happens at many levels. But you've probably met it already. If you have ever looked down into a valley filled with a flat layer of mist or seen the top of a storm cloud spread out, you were looking at an inversion.

CHAPTER 3

The Talk of the Skies

The Seven Golden Patterns • The Cloud Families • Cloud Clues •
Look after Lunch • Meeting the Cousins

CLOUDS WANT TO SPEAK. The legendary navigators of the Micronesian Islands in the Pacific Ocean used the expression *kapesani lang* to refer to the traditional skill of forecasting the weather by interpreting the shapes and colors of clouds. Literally translated, it means "the talk of the skies." The clouds are trying to tell us so much, and we'll be spending plenty of time with them, but let us pause first to remember our aim. We're not in the business of collecting names or identifying clouds for the sake of it. We're trying to decipher their language, and that means meeting the clouds in our own systematic way.

First, let us sharpen the tools: our senses and mind.

No two clouds are ever exactly the same shape, but even if they were, they would never appear identical because the atmosphere changes the colors we see. Clouds that appear white to a close observer pick up blue tones with distance as their light passes through clean air, but yellows, oranges, and reds if there is dust in the atmosphere. The sun casts shadows within and on clouds that change by the second.

Many cloud spotters enjoy finding shapes in clouds, and there is a hidden benefit in this exercise for sign hunters. It is surprisingly difficult to track changes in the shape of clouds: They morph at a rate that is quick enough to be significant, yet too slow for us to follow easily. If we make a habit of noticing a hare, a frog, or some other recognizable shape in the sky, however, the change becomes apparent as the features of the creature flex. If the hare's ears grow, bad weather is likely, for reasons we will explore later in this chapter.

The aim is not to stop and stare but to keep the senses sharp. And we do this by keeping our mind alert with surprises. Our brain quickly grows accustomed to searching for shapes in clouds, and this is the moment to ask it to find shapes in the gaps between clouds. They are there, as strong and distinct as the clouds' forms. On days when there are no recognizable animals in the clouds, you may spot a small zoo in the shapes of the blue gaps—but only if you remind yourself to look for them. They will happily sneak past you if you don't.

Now we are ready to start spotting the signs within the clouds. Very soon we will receive their messages, as the nineteenth-century naturalist Richard Jefferies did: "Dark patches of cloud—spots of ink on the sky, the 'messengers'—go drifting by; and after them will follow the water-carriers, harnessed to the south and west winds, drilling the long rows of rain like seed into the earth."

THE SEVEN GOLDEN PATTERNS

It's time to learn the fundamental patterns of change. They are the nearest we come to universal signs with clouds.

There are a few changes that everyone still recognizes. For example, darkening clouds mean a worsening forecast; the logic is simple—dark clouds hold more water. But even this pattern can be honed, once we have tuned our senses. With practice, we learn to differentiate between a shadow in a cloud and a grey shape that warns of rain.

Many patterns now slip past most people, but they're easy to spot, and each contains a simple message. I call them the Seven Golden Patterns. They apply around the world in almost all kinds of weather and regardless of the exact cloud types. We will look at them in the order of how much warning they give, from long to short. Some may appear obvious, others much less so, but they are all straightforward to use. And it is always worth reminding ourselves that few people notice the obvious in nature.

1. *When clouds get lower, bad weather is more likely.*
This sign grows more useful with our awareness. Many people notice a low, heavy, leaden sky, but far fewer realize it has been creeping lower for hours, sometimes days. This is about tuning in to trends. Consider the following traditional saying: "When the clouds are upon the hills, they'll come down by the mills." This makes more sense in light of the trend than the height of the clouds. It is the fact that the clouds are coming from high to low that is significant, not that they have touched the hills.

2. *The more different cloud types you can spot, the worse the forecast.*
If we see a lot of different cloud types, it guarantees that the atmosphere is unstable at some levels, which increases the likelihood of bad weather. At this stage we aren't trying to identify cloud types or to name them, just to recognize that there are different types out there.

3. *When small clouds grow, the forecast gets worse.*

It sounds so obvious, but most people don't spot this. The casual weather observer notices when the sky has become more cloudy than clear, but not that the small clouds have been growing for hours. The opposite is also true: When clouds shrink, the forecast is improving. We will learn to refine this in many ways, but it is a general pattern that is worth picking up as early as we can.

4. *Clouds that are much taller than they are wide indicate that bad weather is likely.*

A very simple pattern that carries an equally simple but powerful message about instability in the atmosphere.

5. *Spiky or jagged cloud tops are a warning sign of unsettled weather.*

The shape of the tops of clouds is a map that shows us what the air is doing, and pointed shapes or any sharpness mean that unsettled weather is more likely.

The last two patterns lie behind the following lore:

When clouds appear like rocks or towers,
Earth's refreshed by frequent showers.

It refers to the overall shape of the cloud and the rugged texture, especially near the top. By the same token, smooth, rounded cloud tops are a more positive sign.

6. *The rougher the cloud base, the more likely rain becomes.*

The base of clouds tells us whether rain is imminent. If a cloud has a smooth, flat base, it is not a rain cloud.

7. *The lower the cloud we use, the shorter the forecast.*
Low clouds can only reveal what is just about to happen. As we will see, if they start low and grow significantly taller, that is different, but by then they will have reached greater heights and don't count as low clouds.

In these patterns, there was a loose progression from long to short forecasts. A lowering cloud base can give you as much as two days' warning of bad weather, but noticing the rough base of a dark cloud may give you as little as a few minutes. We will delve much deeper into the science and detail of each pattern in stages, starting as we meet the cloud families.

THE CLOUD FAMILIES

If you let them, an official from the United Nations' World Meteorological Organization will slide the International Cloud Atlas under your nose and tell you that there are more than a hundred types of clouds. They will then walk out into the rain, having forgotten their umbrella. This much we know, because before studying the sky, we were all students of human behavior. In every field of human knowledge there are those who enjoy naming things, classifying things, and creating tables. They are good folk, and it is not their fault that they feel naked without a list. Here is a fundamental point I come back to again and again: *If you can't name something in nature, you can still read and understand it.*

This applies to all nature—plants, animals, rocks, the sky. . . . The size, shape, color, and patterns in clouds are trying to tell you stories that the cloud's name can only dream of. Humans were gleaning meaning from the clouds for thousands of years before clouds had any formal names. And there is no such thing as the "right" name for something in nature. Certain cultures accept certain dogmas, but others reject them. Latin names unite some in the West but might be meaningless to an indigenous culture on the other side of the globe. So, as we meet our first clouds, remember that you only need to recognize these general cloud forms: Nobody is going to test you on their names, Latin or otherwise.

To start, we only need to learn to recognize the shape and appearance of three main families of clouds.

The Cirrus Family: high, wispy clouds

Cirrus clouds are the highest clouds we see regularly. Because of their altitude, cirrus clouds are always ice crystals, and this gives them a pure white color. They have many shapes but almost always appear as a collection of thin, wispy strands. They can look like white cotton candy, feathers, scratches, or hairs. Their height makes them appear to be stationary or moving very slowly, but this is relative: They are actually moving fast.

The final golden pattern told us that the lower the clouds, the shorter the forecast. The opposite also holds true: Cirrus clouds offer some of our earliest warnings of change.

Cirrus clouds are joined up high by other members of the "cirro" family. Any cloud with the prefix cirro- is a high, icy cloud.

The Stratus Family: layered clouds

Stratus is Latin for "flat" or "layered," and this is the defining feature of the stratus family of clouds: wide, flat sheets. They can bring rain but more often don't, but whatever they bring will have constancy. And this is the first sign that the stratus clouds offer: no change for a while. If the sky in the direction that the weather is arriving from is filled with wide, flat stratus clouds, there will be little change for hours, and it will be gradual when it comes.

The flat nature of stratus indicates a stable atmosphere.

*The Cumulus Family: heaped clouds**

Cumulus clouds come in several forms. In their smallest, kindest guise, they are best known as the fluffy white sheep of fair weather. On their meaner days, they can grow to alarming towers. Whatever their exact shape and size, cumulus are individual clouds that have well-defined bulges above flatter bottoms. If you see a silken white sack of balls dumped on a glass ceiling in the sky, you are looking at a cumulus cloud. If you have ever watched the opening sequence of the *The Simpsons*, or seen clouds against a blue sky in other cartoons, you were looking at cumulus clouds.

The key to understanding cumulus clouds is to recognize that they are bubbling up. But what does that shape signify?

All cumulus clouds form as a result of warm air rising through convection because of local heating from below. This is an absolutely critical point. It doesn't matter if you're looking at a tiny picnic-friendly marshmallow or a towering giant that looks intent on causing trouble: All cumulus clouds indicate that something local has caused the air to warm and rise vigorously. The rounded bulges at the top are a sign that the air is still rising.

Stratus clouds and cirrus clouds have their own signs, which we will meet later, but in this chapter we will stay with cumulus clouds and use them to sharpen our focus and skills.

---

* In this book, I use "cumulus" to refer to the family of heaped clouds. It's a more concise and elegant word than "cumuliform." Arguably, the best word is the adjective "cumulous," but it's falling out of use, and I won't try to buck the trend. Wherever you see "cumulus" please read it as shorthand for "the cumuliform family of clouds."

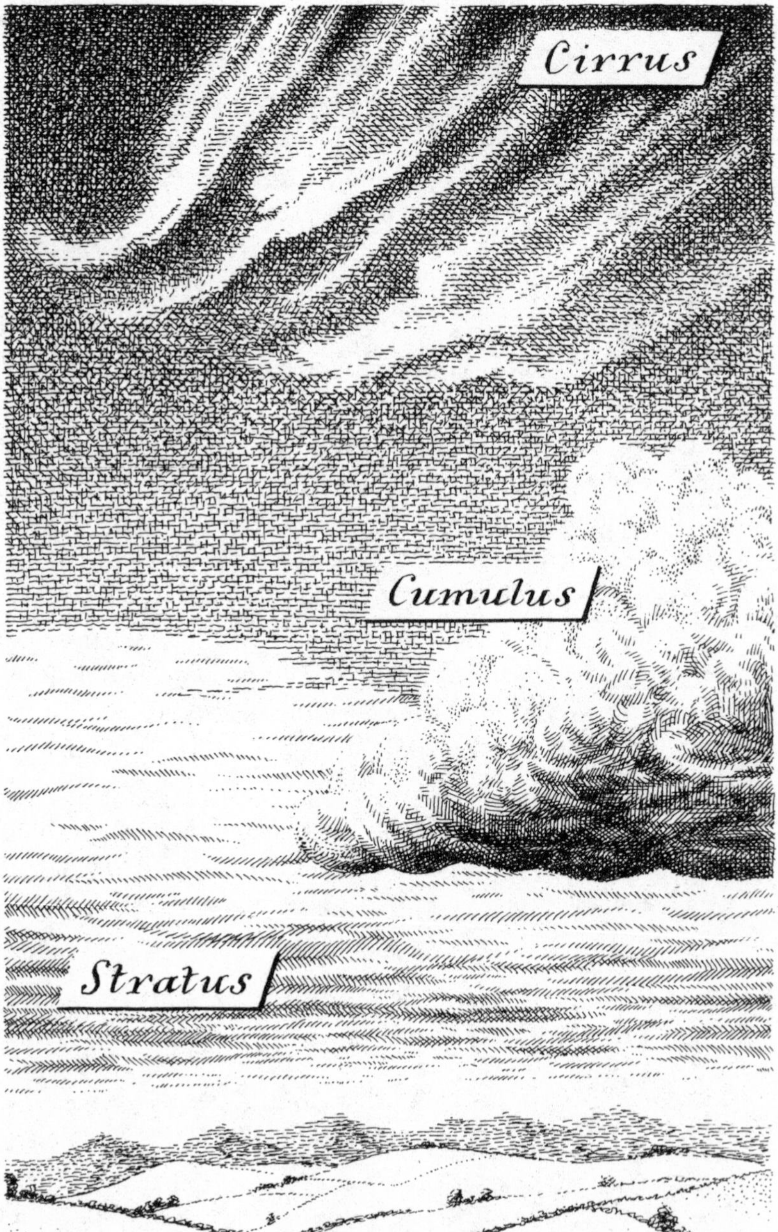
Cirrus
Cumulus
Stratus

CLOUD CLUES

Cumulus clouds draw a picture that reveals a lot about the air, but it also joins the sky to the land.

Every cumulus cloud tells us that there is some moisture in the air, and the size of the clouds gives us a clue as to the level. But it is actually the height of the clouds, specifically the base, that gives us the most useful information.

The more humid the air, the lower the base of the clouds. This means that the height of these clouds measures the humidity of the air for us—it acts as "hygrometer." This instrument has to be read upside down: The cloud base drops as humidity rises. Clouds are lower over oceans than over land because the air is more humid.

Rising humidity is a sign of worsening weather, but humidity is hard to see directly. Fortunately, the clouds bring the pieces together. Lowering clouds mean rising humidity, which is why they signal worse weather to come.

Cumulus clouds tell us that the air is not stable. If the air were perfectly stable, it would not be rising in this lively way. The size and shape of the cumulus is giving us a good map of how unstable the air is. Now we see why clouds that are much taller than they are wide are significant. The taller the clouds, the less stable the air, and the less stable the air, the worse the forecast.

Cumulus clouds are always local, by which I mean they are a result of local conditions. They offer us insights into what is going on in our neck of the woods, sometimes literally. But first we need to ask these clouds the right questions. Since all cumulus clouds are formed by convection as a result of warming from below—thermals—the first key question is: What caused the warming that gave birth to that particular cloud? And there is a good chance that we can point a finger at the culprit.

At one extreme, the low sun never heats the white land of Antarctica sufficiently to create local thermals. There are rarely any cumulus clouds in Antarctica. At the other extreme, the warm, moist air of the tropics means that massive cumulus clouds are a daily event. In the temperate zones, where many of us live, there is an art to figuring out why a cumulus cloud appeared where it did. The answer is always a combination of time and place.

The sun has to give the land enough heat to create the thermal, so cumulus clouds form most easily from the middle of the day until the midafternoon. And we know that the sun does not heat all surfaces equally, so cumulus clouds are more likely over places where there are big differences, like the meeting point of a warm, dry area and a moist, cool one. On a still day, you're more likely to see cumulus clouds over a south-facing wooded hillside than over a lake on a valley floor or even the wood on the opposite, north-facing slope.

The sun has warmed the dark land and woodlands faster than the pale land or sea. This is where we see cumulus clouds bubble up.

It is now entirely up to us how much we wish to refine this art. Sand and grassland reflect much more sunlight than woodlands, so they warm more slowly, but no two terrains heat identically. To start with, when it comes to cumulus clouds, we might want to keep two basic principles in mind:

1. *Nothing is random: Something caused that cloud to be there, and we know that there is local convection—a thermal—directly below it.*

2. *We can often discover what caused that local convection by studying the sun's relationship to the landscape.*

Cumulus clouds can be seen as the icing on the cake of the thermal below them, so our reading of these clouds is intertwined with our ability to see invisible thermals. In the words of the paraglider Bob Drury,

> Thermal detection is a Zen-like art that brings together all your senses with your understanding of fluid dynamics and meteorology. Understanding how the sun, air, and landscape interact to create lift is only half the story. Awareness of all aspects of your glider's movement and position in the air, both vertically and horizontally, is also crucial to build the mental picture we need to visualize the air around us. Our instrumentation should merely confirm what our senses are telling us.

LOOK AFTER LUNCH

When it comes to reading clouds, morning and afternoon are not the same.

On a clear morning we can watch the cumulus clouds trying—and failing, after a valiant effort—to form. Small, uneven, fragmented patches of cloud form and dissipate starting in midmorning. We are watching the system in balance: There is just enough heating to cause a weak thermal and just enough moisture in the air for a cloud to start to form, but neither the convection nor the moisture is sufficient to reach the next stage. It is well worth studying the exact spots where these flimsy clouds appear. If the wind is weak, you will be able to draw a line to the patch of land that is heating more quickly than its neighbors.

As the day progresses, the balance tips. The sun is now giving the land plenty of warmth, and the thermals kick in. The air may be no moister than it was at the start of the day, but it is now being lifted to greater heights and lower air temperatures. The dew point is reached, and a proper cumulus cloud starts to form. The condensing water vapor in the nascent cloud releases heat, which causes more lift, giving the air another nudge upward. Over a couple of hours we have reached a small tipping point, from no cumulus clouds at sunrise, through weak vanishing fragments, to bright white bulges in the sky.

This progression demonstrates why all cumulus clouds are a sign of instability in the atmosphere: They map the places where air is warmed under cooler air, then rises through it. Cumulus clouds will never form without this upheaval, and the taller the clouds, the greater the instability.

The sun's role in this is critical, and it is also the most regular, dependable, and predictable. The variable characters of the air and landscape are where we find more intriguing clues, and it is the air that we should focus on now.

If the air is very dry, we will see no low clouds at any point during the day. The sun reigns. If the air is extremely humid, it is unlikely that we will see much of the sun. But most days that start fine fall somewhere between the extremes, and the key to understanding what lies ahead is spotting where we are on this spectrum. We do this by studying the behavior of the cumulus clouds especially carefully in the second half of the day.

As the sun's heating intensifies, we will see cumulus clouds appear and grow a little. But from midafternoon on, the sun's power is failing fast and the thermals will start to weaken. The million-raindrop question is this: Do the clouds weaken with the sun's heating? If the clouds are being powered only by the convection of the thermals, then as soon as the sun drops and this engine powers down, they will dissipate by losing their crisp, well-defined edges, breaking up and dissolving. If they do, fair weather is likely for the evening and probably beyond. If the clouds do not stop growing as the sun lowers in the sky, bad weather is very likely. If the clouds continue to grow without thermals below them, it is a sign that the air is very unstable. There is a lot of moisture in it, and as this condenses and releases the latent heat, it drives the clouds' growth. On any day that starts with blue skies, we should be extra vigilant in the middle of the afternoon: The cumulus clouds will tell us what the atmosphere—and therefore the weather—hold for us, but only if we tune in.

MEETING THE COUSINS

The three broad families above—cirrus, stratus, and cumulus—cover the broad anatomy of every cloud you will see. Now we should be aiming to recognize this aspect of every cloud's character. Again, we are not trying to pin a perfect name on them, only to make a mental note that a cloud is "very high and wispy," a "flat layer," or "bubbling upward."

Some clouds share attributes of two of the broader families above—they check more than one box. I think of them as the cousins.

Cirrostratus

Cirrostratus is high, like all *cirro-* clouds, but looks distinct and different from cirrus. As the "stratus" part suggests, this cloud spreads over wide areas. But unlike the normal stratus clouds, which are much lower and totally opaque, cirrostratus covers a blue sky as a milky high veil, imperceptibly thin at first.

It is always possible to see through cirrostratus—it rarely hides the sun, the moon, or even the stars very well. Like all *cirro-* clouds, it is high enough that it is made entirely of ice crystals, which can play with the light of the sun and the moon, creating halos—circles of light with a bright sphere at the center.

Halos aside, cirrostratus, being both high and translucent, is the sort of cloud that is rarely commented upon or even spotted, unless we choose to look for it. It is worth the effort to pick out, though, because it offers much in return. Logically, it is a sign of moisture high in the atmosphere, and this, as we shall see, taken with other signs, can be a useful indicator of change on its way, usually things getting worse.

In terms of weather signs, cirrostratus is the most humble, modest cloud. It offers much to those who take the time to get to know and look for it, but it passes over most people without their choosing to take that time.

Altostratus

You will have deduced from the stratus part of this cloud's name that this is another flat blanket. The prefix *alto-* indicates a middle-height cloud that sits between the high cirrus and the low, friendly cumulus clouds. Altostratus is a wide, often thick, opaque rug of a cloud. It can cover a small country.

It sometimes reflects rich colors at the beginning or end of the day, but it's not known for its beauty. In the whole history of weather watching, I doubt anyone has felt rapture, lost in the wonder of nature, when looking at the shape of altostratus. However, sign readers come closest to that state because we see the part it plays in the broader cast of characters. It is much thicker and lower than cirrostratus, so when it follows that cloud it gives us two of the golden patterns: Clouds are growing and getting lower. Worse weather is on the way.

Nimbostratus

Nimbo- comes from *nimbus*, the Latin word for "rain," and nimbostratus is simply a stratus cloud that is rain bearing. It is the least cheery cloud in the sky, a dark grey duvet of dreariness. If it's been raining for half an hour continuously, you're probably under nimbostratus, and because it is a stratus cloud, it will stretch for many miles. There is little prospect of things improving in the next half hour, either.

Cumulonimbus

This is the cloud that everybody recognizes as trouble, but rarely as soon as they might. As the cumulus part affirms, it is a heaped cloud, and as the nimbus part adds, it bears rain.

The cumulonimbus is the storm cloud. It is the turbulent dénouement of an experiment in what happens when air is so unstable it runs riot. Warm, wet air rises through cooler air, lots of vapor condenses, and heat is released much faster than the expanding cloud can lose. It is a cloud with an upward heat accelerator that overpowers the cooling brakes of expansion.

The cloud towers up until finally gravity takes over, as growing lumps of ice, water, and air rush down and up in this big engine of trouble. Friction leads to electrical charges, then bang and rumble—lightning and thunder. Better the devil you know, and later we'll take this troublemaker to one side of the party and get to know him better.

CHAPTER 4

Who Changed the Air?

The Moist Moderator • The Fronts • The Warm Front •
The Warm Sector • The Cold Front • Red Sky at Night, Shepherds' Delight •
Red Sky at Morning, Shepherds Take Warning

A FEW YEARS AGO, I was walking solo in the Lake District, a mountainous region in northwest England. The day started, as it so often does in those parts, with a steep ascent. The rising sun warmed the rocks and air, and soon I was wiping my brow. An hour later, the part of my T-shirt that was sandwiched between my pack and my back was wet with sweat.

About four hours later I was squinting into a freezing fog, brushing ice out of my beard with gloved hands. I had climbed only a couple of thousand feet; the altitude on its own could explain a drop of around 7°F (4°C). Something else had caused the sudden fall in temperature—the air itself had changed.

Throughout much of this book we will be delving into signs nearer the ground—the secret world—but these have to be seen against the theater of large weather changes. Some of the biggest fluctuations we will ever experience come as a result of major changes in the character of the air. They form the backdrop to the secret world.

The main characteristics of air are its temperature and moisture levels. When any large region of air shares the same humidity and temperature, we call it an "air mass." It may be warm and wet, cold and wet, warm and dry, or cold and dry. The character of all air masses is determined by its history and heritage. Ninety percent of the water vapor in the air comes from the oceans, the remaining ten percent from plants, rivers, lakes, and other sources on land. It follows that air that starts life over an ocean will be wet. Air that arrives from a tropical ocean will be warm and wet, and air from a polar land mass will be cold and dry. Air masses can change a little over time: They can get drier or wetter by passing over land or sea but, perhaps surprisingly, they don't mix well. That is why we experience such sudden changes.

In October 2019 Denver, Colorado, experienced a drop in temperature from 82°F to 28°F (28°C to −2°C) in a day. However abrupt that may feel for residents of Denver, in weather terms it's plain: A warm air mass was bumped out of the way by a cold one.

Each air mass has its own temperature profile. The way in which the temperature changes with altitude is different in each air mass, and this will determine how stable that air is. Polar air tends to be cold and stable, while tropical air is typically warm and unstable. The different moisture and stability levels of each air mass help explain why not all sunny days are the same. The Wola people of Papua New Guinea are familiar with the sort of humid unstable air that quickly turns to heavy rain and storms as soon as the sun warms the land and sets up thermals. *Chay nat*, literally "rain sun," is their succinct and elegant expression for "warm, wet, and unstable air mass."

In the US and UK we may encounter settled dry weather during summer high-pressure systems. The air doesn't feel humid or muggy, just dry: the "dry high."

Each part of the world has its own set of common visiting air masses, determined by the neighboring areas of land or sea on each side. If you are in a part of the world totally surrounded by land or sea for hundreds of miles in all directions, you can expect long periods of similar weather. Oceanic islands and landlocked regions of Europe, Asia, Africa, Australia, and the Americas regularly go weeks without the weather appearing to change dramatically. Constant weather is much rarer in the UK and anywhere else in the world that is surrounded by very different air masses on each side.

Air masses determine extremes of temperature in summer and winter. Air that arrives from over land can mean very hot weather in summer or bitter cold in winter, but air from the sea means milder temperatures in all seasons. Wind direction can give you a good idea where the air has come from. Notice how extremes in summer or winter so often coincide with winds that blow from land, not sea. In the US, the coldest snaps tend to arrive from the north, with a cold air mass that passes through Canada; in the UK, I'll usually feel the most frigid winds blowing from the east, courtesy of air traveling from the Continent.

When it is sunny early in the day, we will want to get a feel for the air—is it muggy or dry? Is this a humid, unstable day or a dry stable one—are we seeing a "rain sun" or a "dry high"? A rough guess is all we're after at this stage.

If the summer air feels muggy and humid and we sense a rain sun, we can predict a serious deterioration in the weather later in the day. The small clouds will grow and keep growing, leading to possible storms. But a dry high will lead us to expect more settled weather: The cumulus puffs will grow after lunch but soon start shrinking again.

THE MOIST MODERATOR

Land temperatures fluctuate much more rapidly than sea temperatures. The sea has a temperature that is hard to budge—it can take weeks for an ocean to gain or lose a few degrees, while the land may take hours. This means that islands and coasts share the moderating effect of the sea: Their temperatures are less extreme, making them milder in winter and summer. I call this effect the "moist moderator," and it's worth noting if you're anywhere in the world where the weather comes in from the sea. You can expect it to be damper and milder than it is even an hour's drive inland. If there is any snow on a beach, it will be deep not far away.

The oceans take a long time to warm, but they take their time to cool, too: They store heat well. Oceans act like heat repositories, and their currents transport heat around the world, which helps explain why many places in the world at the same latitude have totally different climates, like Edinburgh and Moscow. These currents have long tentacles: Those off Peru can affect Australia, a phenomenon known as El Niño.

THE FRONTS

Early one morning last week, all bundled up, I walked into a field and took in the sky. The weather was fine, but the signs were loud and the outlook was bad. The sky was mostly clear, but with no pure blue anywhere: Cirrus and cirrostratus shared the highest levels. The low and high clouds moved in different directions. I knew that rain would fall before I went to bed that night.

Later that day I led a group of walkers on a natural navigation walk around the Weald & Downland Living Museum in Sussex. Two people mentioned the weather forecast, but nobody commented on the weather. I silently marveled that the sky was not a topic of conversation for our first hour together. A barn dance of cloud and wind changes was taking place above our heads. By the end of our walk, everyone was talking about the sky. It had metamorphosed to the point where it could no longer be ignored. We felt the first raindrops as we stopped in the late afternoon.

This is a cultural pattern you will spot, too. People talk about forecasts, throwing around words such as "apparently," "on its way," or "sounds like," but few notice the clues that major change is on its way.

We'll bring those clues back into focus and give them their voice again. There are practical reasons for doing this, but aesthetic ones, too. The signs that herald rain are more beautiful than the rain itself. The keys to spotting them lie in the "fronts."

Whenever the air mass we are in is replaced by another, we notice a significant change in the weather. The boundary between two different air masses is called a "front." When forecasters say, "A front is about to go through," they mean, "The air mass we are currently in is about to be pushed out of the way by another, and there will be a noticeable change in the weather."

Fronts are named for the temperature of the air they bring in. When a warm air mass replaces a cold air mass, it's a warm front, and vice versa.

There are signs that precede, accompany, and follow each front going through. For forecasting purposes, the signs that precede a front are most valuable, but there is a progression as one front follows another, so the signs that follow one can be used to predict what will come next.

These roaming packages of warm and cool air are better known as "low-pressure systems" or "depressions," because the air pressure drops toward a low at the center of the system. It drops noticeably as they approach, which is why ships have carried barometers for centuries. We can't sense these pressure changes directly, so we'll focus on the moment when the air masses change: the fronts.

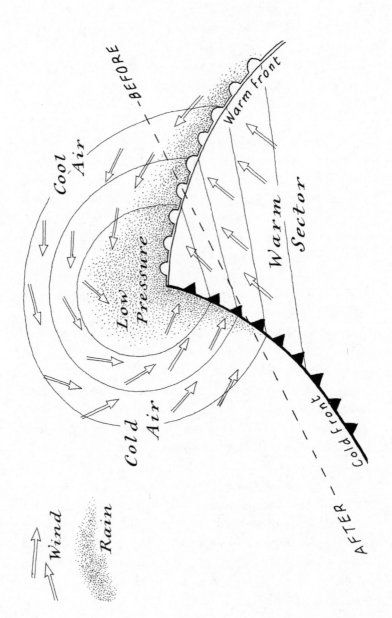

Wind

Rain

Cool Air

BEFORE

Warm front

Low Pressure

Warm Sector

Cold Air

AFTER

Cold front

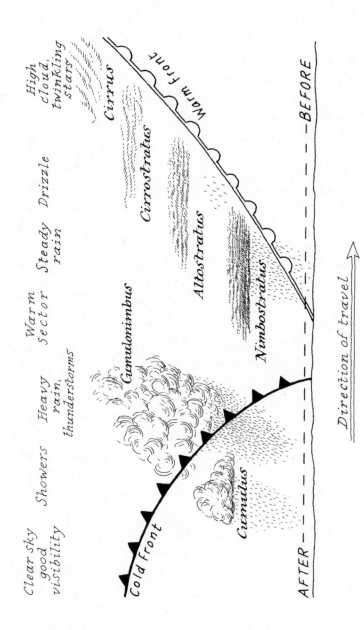

Cold Front

Clear sky good visibility

Showers

Heavy rain, thunderstorms

Cumulonimbus

Cumulus

Warm Sector

Steady rain Drizzle

High cloud, twinkling stars

Cirrus

Cirrostratus

Altostratus

Nimbostratus

Warm Front

Direction of travel

BEFORE

AFTER

A period of fine weather is often broken when a warm front and then a cold front go through in succession. The warm front brings a segment of warm air—known as the warm sector—followed by cold at the cold front. The typical temperature changes as the weather breaks: getting warmer, remaining warm for a while, then colder.

Let's look at these three stages—the warm front, the warm sector, and the cold front—in a bit more detail.

The Warm Front

The arriving warm air is less dense than cold air and is forced up by the cold air. It then slides a long way over the top of the cold air. As the warm air is forced up, it is cooled, leading to condensation and clouds. This is helpful because some signs reach far ahead of the main weather change over clear skies, making them easy to spot.

Starting with the blank canvas of perfect blue skies, one of the first signs that a warm front is approaching is found in the highest, wispiest cirrus clouds. These thin streaks often show a marked direction trend in the sky—from west to east is common—and reveal the direction from which the front is coming. The cirrus marks the leading edge of the new air mass and indicates that bad weather is twelve to twenty-four hours away. It is hard to be more precise using just this sign because each front travels at its own speed. Even so, the leading-edge cirrus is a gem: The shallow gradient of the warm front means that its top may be as much as 930 miles (1,500 km) ahead of the front at ground level. Few signs in nature let you spot something coming from as far away as Rome is from London.

You will recall that warm air over cold air is very stable because cold air will not rise through it. As this warm, moister air slides over the colder air, it creates an inversion, capping the air below it. If there are any tall cumulus clouds among the blue skies, you may notice that they appear stunted near their tops as the warmer air sits on the cooler.

A steady progression of clouds follows. The cirrus builds from a few lines to a well-clawed sky and is replaced by the frosted glass of the high cirrostratus. The sun or moon are still visible, but they appear muted, and halos are common around both. Now a wedge of warm air is sliding over you and the clouds drop lower and thicken. Next, a blanket of altostratus blocks out the sun or moon, and the sky is opaque. The bad weather is getting nearer: Rain may be only four hours away.

The wind is now stronger and has backed—meaning it has shifted to be coming from a more counterclockwise direction: If it had been coming pretty much from the west before, it will be arriving from more to the south now. The clouds are lower still and growing darker. The nimbostratus arrives, and with it the rain. Visibility worsens. The rain passes over in bands of heavy spells interspersed with weaker patches or drizzle.

The front is now ready to pass. The rain continues, but the wind veers, shifting clockwise. Warm-front rain is steady, dreary, determined, and dull. It usually lasts several hours but is more unpleasant than unsafe. The four words to sum up warm fronts: gradual worsening to wetness.

In the early days of learning to recognize a front going through, I encourage cheating: Lean on the meteorologists and their charts. Wait until you have been enjoying a period of very fine weather, then keep an eye on the forecasts. Keep doing this until you hear or see that a warm front is about to go through in the next thirty-six hours (on weather charts a warm front is marked as a dark line with red semicircles on the leading side).

Now you are ready to spot the progress of the signs. To recap, when a warm front goes through, keep an eye out for the following features:

- leading-edge cirrus
- "frosted glass" of cirrostratus
- halo around the sun or moon
- stunted cumulus clouds
- sun disappearing as altostratus thickens
- wind strengthening and changing direction, normally backing
- lowering clouds
- darkening clouds
- bands of rain
- visibility deteriorating

The Warm Sector

Once the warm front has passed, we're in the middle of the frontal sandwich. There is a warm front ahead, a cold front still to come, and we're sitting in the warm sector. The skies have a blanket of mixed clouds, usually of a stratus nature, and the air is warmer than it was before the front passed. There may be light rain, drizzle, or none. The winds are steady and visibility is poor; fogs are common.

The warm sector is a damp, mild interlude between the more substantial events of the two fronts.

The Cold Front

Warm fronts spoil us, as they give us so much warning and approach over good weather and visibility. Cold fronts normally arrive after the deterioration and bad visibility wrought by the warm front, making them harder to track. But that is only the start of the challenge.

The cold air slides under the warm air ahead of it, so instead of our watching a leading edge far above and ahead of the change, cold fronts arrive suddenly. This is a shame, as the weather they bring is more riotous. Cold fronts sneak up on us and arrive with a bang, sometimes literally.

The reason cold fronts bring more dramatic weather than warm fronts is that they slide under the warmer air ahead as a steep wedge. This forces the warm air sharply upward and triggers massive volatility. It also means that the weather at ground level changes at almost the same time as the weather at high altitudes, so we can't use the long lead time of high clouds to spot cold fronts in the way we did with the warm front.

The poor visibility before the cold front arrives can make cloud spotting hard, but you may see many clouds that are taller than they are wide, which you are unlikely to see associated with either the warm front or the warm sector. As the front approaches, the wind strengthens noticeably and changes direction markedly, normally veering sharply before, then backing at the front. There is a sudden drop in temperature at the front.

Instead of the leisurely layers of dampness and well-behaved clouds that the warm front brings, the cold front ushers in a wave of upheaval that sometimes includes violent weather and storms. However, the steep wedge of cold air that creates the drama also offers the cold front's only small mercy: The mayhem is short-lived. Soon after the very bad weather arrives, it passes to reveal cool air and much clearer skies: the "clear slot." The air is still unstable, so you might see scattered cumulus or cumulonimbus clouds, with some showers lingering, but visibility is suddenly much improved.

When you suspect a cold front is approaching, keep an eye out for:

- strengthening wind that veers, then backs
- sharp drop in temperature
- very unsettled weather, tall cumulus clouds, and possible storms
- a patch of clear skies and cold air soon after

The fast passage of cold fronts is probably the logic behind the old saying, attributed to Charles Wesley, the eighteenth-century Methodist and prolific hymn writer, "Sharper the blast, sooner 'tis past."

The signs above mark the progression of standard fronts. Each one you see will share enough of these features that you will be able to recognize it, but each will have its own character. There are also local trends, and the fronts in each part of the world may demonstrate their own idiosyncrasies. Cold fronts in central North America are often preceded by southerly winds and can lead to strong snaking winds at night.

I should also let you know about a situation that might trip you up if you were unaware of it. Cold fronts travel faster than warm, and eventually catch up and merge with them. This leads to an "occluded front," where all of the weather and signs above can appear to arrive at almost the same time. We experience a lot of bad weather apparently all jumbled together, and the sign reading gets harder.

At this stage don't worry about details: We just want to be comfortable with recognizing the broader characters of the warm front, warm sector, and cold front. There is no need to memorize all the signs—we're not trying to remember what color shoes they'll be wearing: We just need to recognize their big fat forms as they barge into a room.

RED SKY AT NIGHT, SHEPHERDS' DELIGHT

A red sky at night tells us just one concrete thing: The sky is clear to the west, allowing the sun's light through. The most impressive red skies show the low sun bouncing off clouds in the east, probably having passed us already. Since most of our weather comes from the west, the short-term forecast is good.

If a cold front has gone through during the afternoon, a red sky in the evening is likely, and cooler, fair weather can be expected. You'll be able to decide if that is the case by looking for the passage of signs above.

RED SKY AT MORNING, SHEPHERDS TAKE WARNING

If the weather has been fair but a front is near in the morning, you may see the eastern sun lighting up the approaching clouds in the west. The forecast is bad. You can refine it by looking at the clouds: If cirrostratus is followed by the lower, thicker altostratus, then the shepherds you spoke to were right.

How to Feel the Wind

THE PATTERNS IN THE PUDDLE pinned me to the spot. Gusts of wind ruffled the surface of the shallow water, but during the many lulls I could see a rich picture in the water mirror. There was a wind map in the reflections.

The sky was mostly blue but for a few scattered cumulus clouds and, between them, high clouds, cirrus, just visible in the gaps. On the other side of the puddle, and well below any clouds, there was a pair of leafless silver birches. Standing by the puddle, I could sense the wind at five levels. The high clouds were being blown from northwest to southeast. Below them, the cumulus clouds were moving from west to east. A similar wind ruffled the tops of the birches. Over a few minutes I could feel a breeze on my face from more than one direction, and occasionally a strong gust would reach down to send ripples across the puddle, sometimes from a different direction from the one I felt on my face or saw in the clouds. So, what was the wind direction?

We need to simplify. We'll refer to three wind levels only: high, main, and ground.

The high winds pass over the top of our landscapes and much of the weather we feel. We sense them mainly by looking at the highest clouds, like cirrus.

The main winds are affected by the landscape they blow over, but not dominated by it. The greater the altitude, the stronger these winds are and the less their direction is influenced by the ground. We sense them by looking at all levels from treetops to all but the highest clouds.

Ground winds are strongly shaped by their relationship to the local landscape. These are the winds we feel. Most meteorologists are not very interested in them, as they are too local and variable for regional forecasts.

Almost all wind measurements and forecasts refer to main winds, not ground winds. If we want to interpret the signs around us using our senses, we need to tune in to ground winds, since we spend our lives in this lowest layer. Fortunately, this band is full of great characters, and many are rebellious. They inhabit the secret world, and we will meet them soon.

When we refer to "wind direction" in a general sense, we're referring to the main winds. If we're sensing them directly at ground level, they're ground winds. If we're discussing the highest clouds, they're high winds.

It's time to fine-tune our senses once more by considering three signs, one for the ears, one for the eyes, and one for our skin.

THE WHISPERING FORECAST

We can hear the wind, which is strange because the air itself makes no sound. The wind makes a sound only when moving air interacts with something else. So when we hear the wind, we can ask: What is the wind wrestling with to create that sound?

It is usually easy to find the cause, but it's always worth pausing to consider. The wind makes sounds in several revealing ways.

It sets some objects moving, making them roll, flex, and bump into each other or along the ground, like leaves scratching and skidding along stone. Willow trees make a telltale creak in strong winds as their branches resist the bullying. The wind causes things to break or fall. It's a rarer drama, but the next stage for willows that can't bend any further is a shattering, cracking sound. That's how the crack willow (*Salix fragilis*, native to Europe and southern and central Asia but thoroughly naturalized in the US) acquired its name.

The most common way in which wind creates sound is from the vibrations caused by friction. Wind blowing over a surface that is perfectly smooth would be silent. But such surfaces exist only theoretically, so wind of sufficient speed blowing past any surface will generate friction and sound.

The conditions have to be perfect for us to hear something as distinct as a whistle or other crisp note, but we don't have to wait for these rare concerts. We can learn to listen to the wind's daily practice sessions. Every time we hear it, however faint, we can ask two questions: What instrument is being played and how? This little exercise sounds airy and fanciful, but it yields results. It nurtures one sense until it builds a relationship with wind strength and direction, which offers many practical benefits. When the wind passes vegetation, it sounds very different from when it flows over sand: This is why you can hear oases in the desert.

The sound of the wind passing a rock, a hill, a valley, an island, a wood, or a street will change if its strength or direction alters. Any shift in wind direction is often the clearest herald of bigger weather fluctuations to come. Noticing any deviation in the sound of the wind—its pitch, volume, or character—can signal things that our eyes have yet to pick out: the "whispering forecast."

THE WIND DANCE

We pay little attention to the strength of the wind. We may comment on a gale or a storm, but we mostly ignore the nuanced shifts.

As the wind grows from nothing to the very gentlest breeze, something so light we cannot feel it on our faces, a dance has begun and there are changes all around us. Smoke bends, aphids fly, and spiders take off. When it strengthens a little, so that we can just feel it on our faces, we hear the first leaves rustle and see the lightest seeds float by. A touch stronger still and dust is picked up, twigs flex on trees, winged seeds are set in motion, thermals cease, and circling birds disappear. Aphids and spiders are now grounded.

A notch faster, and now we hear it in our clothes; branches sway, and gnats and mosquitoes stop biting. A little stronger: Leaves take off and flies stop flying. If you see large branches moving, the bees and moths have given up. And by the time twigs break, there is debris in the air, we find walking difficult, and most flying creatures have long since taken shelter . . . but, somehow, dragonflies are still aloft. Swifts will stay airborne even as branches break, children are blown over, and all insects have been grounded.

That is a lot to take in. To begin with, take note of the behavior of leaves in the wind. Watch their movement on trees and on lower plants, on dead leaves on the ground and any that have taken to the air. Now notice how the height of grasses gives them their rhythm, and watch how this rhythm changes by the second.

THE COOL, FAIR BREEZE

If a wind feels cooler than we're expecting, this may be a reflection of the air's temperature and speed. But it can be a guide to humidity, too. On cold days, the wind's speed has a much greater impact on us than humidity, but on warm days, the drier a wind is, the cooler it feels. This is because the drier the air, the easier it is for evaporation to take place, and evaporation makes us cooler—that is why we sweat. Since lowering humidity normally accompanies improving weather, this leads to a counterintuitive sign: On warm summer days, a cool, steady breeze is a sign of fair weather to come.

THE NEEDLE AND GAUGE

I remember vividly a journey I made in my mid-twenties. I was driving a cheap old car to an airfield near Maidenhead, Berkshire, for some flight training. Pilots have to follow a routine of checking their "T's and P's"—the temperature and pressure gauges of the engine. The habit was drilled into me to the extent that I automatically checked the temperature gauge of the car I was driving—I couldn't help it.

A few miles from the airfield, the car was traveling smoothly, but the temperature needle started to rise quickly, then went off the scale. I pulled off the road, just as the head gasket blew. Steam and oil smoke erupted from under the hood. The engine was knackered, but the gauge had given me just enough warning to get off a fast road in time.

When the needle on any gauge moves significantly, something has changed. Wind direction is the needle on the weather-engine gauge. If we stay tuned to it, it will warn us of change before it hits us.

Aristotle's successor, Theophrastus of Eresos (c. 371–287 BCE), wrote a book called, tantalizingly, *On Weather Signs*. It includes eighty rain and forty-five wind signs. Like many such ancient texts, it is brilliant and baffling by turn. He overreaches at times, but he is spot on when he notes that there is a strong connection between wind direction and weather changes.

A sensitivity to wind direction is the most underappreciated forecasting tool at our disposal. But why and how does it work? To answer that we need to look at the large systems and major changes, like we did with air masses.

Air is always on the move. It is always trying to equalize by flowing from areas of high to low pressure. When we release the neck of a balloon, the high-pressure air escapes and flows quickly toward the lower pressure in the room.

Wind, the horizontal movement of air, is driven by this equalizing force. Over short distances, wind is as simple as that: An area of higher-pressure air rushes toward an area of lower pressure. Over longer distances it follows a curved path, as a result of the Coriolis effect. The Earth's rotation causes winds to bend to the right in the northern hemisphere and to the left in the southern.

The sun's radiation warms some areas, like the tropics and land, more than others, like the Arctic and the oceans. This causes an imbalance of heating over the Earth. Warm air rises and expands, causing its pressure to fall. Cool air sinks and contracts, causing its pressure to rise. This gives us high- and low-pressure areas: The air over the equator is warm and low-pressure; the air over Antarctica is cold and high-pressure.

Air tries to flow from high to low pressure, is deflected by the Coriolis effect, and ends up circling clockwise around high-pressure systems and counterclockwise around low-pressure systems. This is why, in the northern hemisphere, with our back to the main wind, there is either a low-pressure system to our left or a high-pressure system to our right, or both.

It follows that any significant alteration in the wind direction must herald a change in the layout of high and low systems near us. The pressure systems determine the movement of the air masses, and therefore any major shift in the wind means a significant change is on the way. This is a fundamental point, so let's recap: change in main wind direction = movement of pressure systems = air masses near us moving = major change likely soon.

The Indian wet monsoon is known for the way it brings a vital season of heavy rainfall. It begins when the sun's seasonal journey north changes the high- and low-pressure balance, causing dry air that has been arriving from the landlocked northeast to be replaced by moist air that flows in from the Indian Ocean. The rain is important, but the monsoon is actually a seasonal wind direction, one that blows from the southwest between May and September and from the northeast between October and April. The wind direction and the weather can be thought of as the same phenomenon: When one changes, the other does, too, emphatically.

In drought-sensitive parts of the world, like the southwestern US, people traditionally look for a change in wind direction as the signal that the drought is ending. And we have already seen how a wind change can be one of the signs that a front is about to pass through. The scale varies, but in each case a change in wind direction signals that a different air mass is about to arrive.

The science is more complex than the sign, but now that you know both, you can use the simple sign with confidence:

> A major shift in the wind direction means that a major weather change is coming soon.

GETTING TO KNOW THE NEEDLE AND GAUGE

History is sprinkled with examples of fortune favoring the wind-aware. George Washington kept a detailed weather diary, and it may not be stretching things too far to say that this habit changed history. In January 1777, the winter before Valley Forge, Washington's army was trapped by British troops. On noting that the afternoon's wind was from the northwest, he was able to predict colder weather overnight and subsequent hard, frozen ground. This allowed him to risk an escape route that would have been too soft and muddy in warmer weather.

In all parts of the world, the wind brings the weather, and the regional geography will have its say. Or as Francis Bacon, the English philosopher, had it, "Every wind has its weather." We can think of the winds as a sort of mail, with recognizable envelopes and postmarks. I'm sure we all glance at envelopes for clues as to the nature of the contents before we open them. The marks on the outside of the envelope can reveal the origin of what is inside and hint at what we're about to meet. State or municipal envelopes might warn of taxes or fines, but anything handwritten under a favorite postmark can be opened with optimism or eagerness. (When I was seventeen, I was at an all-boys boarding school and could recognize my first proper girlfriend's letters from about a hundred yards away. These days, the same person emails me to-do lists. It's not the same. But I digress.) The direction of the wind gives us a strong clue to what it is bringing with it.

Across cultures in the northern hemisphere, it's easy to find references to cold winds from the north, but there are regional traditions across the globe. The Bible gives numerous warnings about wind directions, and three Old Testament books—Job, Isaiah, and Zechariah—warn that winds from the south are bad news: "Out of the south cometh the whirlwind." (Job 37:9)

This is why in many regions there is sensitivity not just to a change in wind direction but to any shift toward a certain quarter. In each case an understanding of the relationship among land masses, any high ground, and seas will add to our picture.

The richest history of this habit can be found in the Mediterranean, which has landscapes of different character on all sides, and there the winds have earned their own reputations. Their names will usually be tied to visible or memorable features in the direction they have come from. The tramontane—from the Latin *trans montanus*—means "across the mountains" and blows down from the Alps in the north. Where there is no distinctive terrestrial feature to anchor a wind's name to, celestial objects fill the gaps. The easterly Mediterranean wind is called the levant, which ties it to the sun and stems from the Latin verb "to lift up."

If we think about a wind that changes from south to north, we have noted both the significant change in direction and the quarter that it now comes from—the north. But two pieces of information are missing: the how and the when. As FitzRoy reminds us, the *way* in which the wind changes is significant: "The strongest winds are when it turns from south to north by west." In this case he is noting that a wind that moves clockwise, veering from south to north, means something different from one that backs counterclockwise through the east, even if both start and finish in the same place. This makes sense when we think back to the fronts: A veering wind and a backing wind are telling us different things about what comes next.*

That is why we find this direction of change wrapped up in local lore, such as this old English saying:

A veering wind, fair weather;
A backing wind, foul weather.

---

* Many people who juggle more complex issues in their daily lives find the labels "veering" or "backing" wind confusing at first. For some, it helps to remember that when the wind backs it goes against the clock. If this helps, great. If not, feel free to ignore it.

The Prussian meteorologist Heinrich Wilhelm Dove adds the final ingredient. There is always an important seasonal aspect: "If the wind shifts from south to north through west, there will be, in winter, snow; in spring, sleet; in summer, thunderstorms, after which the air becomes colder."

We now have the four ingredients of a major wind change: both the direction it was and is now coming from, the direction it swung (did it veer or back?), and the time of year.

When starting out, it is worth keeping it as simple as possible. You can consider yourself promoted to the upper echelons of awareness simply by noting that the wind direction has changed— few do these days. Thankfully, it's a quick habit to regain, and there are ways to accelerate this. Once more, I'll encourage cheating.

Whenever you hear a forecast of a major change in the weather, note the current wind direction and monitor it over the coming hours. Even if you're comfortable and familiar with attaching cardinal directions to the wind—noting that you're experiencing an easterly, for example—I would still urge you to give your wind a visual anchor. Pick something in the distance that marks the direction the wind is coming from. If there are any visual gifts, like flags flying or smoke blowing, you can accept these gratefully, too.

Visual anchoring helps in two ways—first, with our memory. When starting out it is all too easy to jumble southwest and southeast in our minds, and a physical object we can see acts as a check on this. Second, it acts as a visual prompt to check in with the wind. It is much harder for a change in wind direction to pass you by if you have it tethered to a tall building or copse on a hill.

If you keep doing this from a familiar spot, it will not be long before you start to notice a coupling. For example, when the wind shifts to come from one landmark like the church steeple to a different one, like a distant summit, rain sets in four hours later. This is something that farmers and rangers, who work the same patch over years, come to recognize.

Once we have clocked these changes in a home area, we can, if we choose, layer the cardinal directions on top of the changes. Instead of thinking of the wind as the summit or church-steeple wind, we label them a southwest and northwest wind. This makes the knowledge and awareness portable: We can carry it with us as we explore an area.

To further refine your reading of wind direction, I'd encourage you to close your eyes. Our eyes are too powerful for their own good and like to override the other senses. The technique I use is to find the highest nearby open ground, close my eyes, and turn my face until I feel the wind equally on both cheeks and, if strong enough, hear it equally in each ear. Then I raise my hand and very slowly karate chop the air, shifting my hand until I feel the wind cooling each side equally. Only then do I open my eyes and pick my landmark in the distance.

This technique keeps your eyes from latching on to something in the rough direction that the wind is coming from: It may be very convenient, but it isn't wholly honest. What this means in practice is that you often notice the wind coming from one side of prominent landmarks. In this case, I form a clenched fist and say out loud, "Three knuckles to the right of the summit." I wear the odd looks I get with pride.

This awareness will stand you in good stead for monitoring the broad wind and weather trends. Now we're ready to lower our focus.

GENTLY DIVERGING WINDS

There will be many times when we look up and see clouds at slightly different heights moving in similar but different directions. This can be confusing until we're comfortable with the way winds circle a low-pressure system and how this changes with height.

Above 1,500 yards, the main wind is not much affected by the ground, but below that level, friction causes it to back and lose speed. The lower the main wind, the more it is slowed and the more it backs, bending to come from a counterclockwise direction.

You can visualize this as the stopper in a sink. Fast water swirls in a circle around the hole, but if anything slows it down, it falls more steeply into the hole. The winds circle the center of a low-pressure system in a counterclockwise direction, but any slowing by friction leads to them to "fall down the drain" of the low. They bend in, left, toward the center. They back.

This is why low- and middle-height clouds move in similar but slightly different directions. The lower set is being carried by winds that have been slowed and bent more by friction.

SMOOTH OR TURBULENT?

After a wave has broken at the beach, the water rushes up the sand in a smooth, flat way. But have you noticed how, as soon as the water hits any obstacles, like pebbles, the smoothness disappears and the water becomes rough and chaotic?

Scientists who study how gases and liquids flow divide their world into two broad behaviors: laminar and turbulent. When a fluid is moving smoothly and in a consistent direction, the flow is said to be laminar. When it starts to whirl and form eddies, it's turbulent. We can think of the wind as having these two worlds, too. Well above tree and building height, the winds flow according to the pressure systems—they follow simple paths in a laminar flow. Below that height, they start to twist, bend, and do all sorts of turbulent things. Main winds tend to be laminar, but ground winds are more likely to be turbulent.

That's a lot of slightly abstract theory to be chewing on, so let's spit it out for a second and return to the main aim, which is identifying the signs to look for. Next time you're in a town or surrounded by trees and see a few clouds pass over your head, make a mental note of the direction the clouds are coming from and also notice how this does not change noticeably over many minutes. Now walk around the block, if you're in town, or around the clump of trees, if you're in a rural spot, and notice how the wind speed and direction you experience on the ground changes a lot. You have just witnessed the wind in laminar mode, at cloud-carrying height, and in turbulent mode, down among obstacles.

CUMULUS EQUALS GUSTS

Now we're ready to watch the wind bounce off invisible obstructions. On the next morning of good sunny weather, with steady, very light winds, find a wide-open area that is not too cluttered with trees or buildings. Use the techniques I mentioned above to get a good sense of the wind direction, but also gauge its strength and character. Try to get a feel for how steady it is. Is the breeze blowing at a fairly constant speed or is it volatile, with a speed that fluctuates? Now repeat the exercise in the midafternoon. Did you notice a difference?

Was the afternoon breeze gustier? It might not be any stronger overall, but its character is much less consistent than that of the morning breeze. To understand why, we will return to convection.

In the morning, the sun has not yet heated the land, so the wind in open country will be governed mainly by pressure differences. But by the afternoon, the sun has warmed the land, which has created thermals, and these pillars of rising air have a similar effect to buildings. The breeze can no longer flow in a straight line: It is being jostled by columns of rising air. This leads to fluctuating wind speed and direction—a gustier wind.

Now we can bring together two related observations to act as simple signs to each other. Thermals create heaped cumulus clouds and lead to gustier wind conditions. So, cumulus clouds on a sunny day are a sign that the wind will be gustier, just as a gustier wind on a sunny day is a prompt to look for cumulus clouds.

The clouds sit on top of the thermals, which is why they can follow steady, straight paths above us, even as we feel the wind shift a little bit every few moments.

SAND AND SANDWICHES

While you're tuning in to a wind's gustiness, something else is worth looking out for. When a wind blows over small particles, like sand, dust, snow, or dry leaf fragments, it sometimes picks these particles up and carries them; at other times it leaves them on the ground untouched. The wind speed is a major consideration, of course—if it's too weak it will lift nothing—but gustiness is also a significant factor. A weak gusty wind can pick up more particles than a stronger laminar wind. I have noticed this with dust in towns, leaves in the woods, and sand on the beach in summer: The strong breeze doesn't always whip the sand into the sandwiches the way the weaker gusty one does.

This is worth knowing if you're picking a spot for beach towels. Most people assume that the sand in the air is there because of the wind speed that day, and there is little anyone can do about that. But it's as much a measure of turbulence, and this is something that changes over tiny distances. The shape of the beach upwind of you is a massive factor: Sometimes you need to walk only a few feet to find that the wind speed feels the same but the sand is no longer in the air. I can't promise there will be no sand in your sandwiches, but they will crunch less if you search out the smoother wind.

SUMMER CITY GUSTS

Alongside strength, direction, temperature, and sound, we always have a new characteristic to consider: turbulence, or gustiness.

Have you noticed how on sunny warm days in cities, even when the wind isn't strong, it has an unusually gusty nature? The convection and buildings combine to create ideal conditions for turbulent breezes, and the gusts will reach more than double the average wind speed in cities. Summer gustiness is an urban wind signature: You won't encounter exactly the same sensation anywhere else.

AMONG THE GROUND WINDS

A few weeks ago I noticed a change in direction of the main, well-above-treetop, cloud-carrying wind. It backed by almost 90 degrees. This was a sign that a front was about to go through: Rain was on its way.

Soon after noticing that major change, I met Hannah Thompson, a ranger from the National Trust. We set out on a walk, and Hannah showed me an ambitious reforesting project near where I live in West Sussex, called the Rise of Northwood. We passed several of the thirteen thousand new saplings, with their protective collars and wooden stakes. Hannah pointed out something that was logical but which I had never previously noticed: The stakes for supporting trees are normally placed in the ground on the side that the prevailing winds come from, the southwest in the UK. Compasses, compasses, everywhere!

We walked down between two older woods and saw a pair of stonechats—a small songbird—on the ground by the path. The wind was on our backs, but it was fluky, shifting, and I felt a gust on my face. Hannah opened the metal deer gate, and we headed into the main area of planting. A kestrel was diving, banking, busy persuading a buzzard to buzz off. The trees were only a few years old but were doing well, except in one area.

As we climbed a gentle hill the young saplings gave up: The ground was marked by withered, sad stems within their stakes and collars. Hannah explained that the trees just would not take to some areas, patches of excessively stony or wet and boggy ground. But this area was not wet or excessively stony—it was not very different from the places where the trees were thriving, yet these were a picture of failure. As we took a few more steps, the cold wind grabbed at the back of my neck and blew away the mystery.

Where Hannah and I now stood was the midpoint between our two woods. The winds were dipping down into this patch, scouring the ground, then leaping up and over the woods to our north. Leaning down to inspect the beaten saplings, I could see their wind-ravaged little leaves pointing to this cause. At that moment, the feel of the wind, the shape of the land and the woods, old and new, were one.

I strongly recommend trying a little experiment the next time you find yourself in an open area between two woods on a windy day. Try standing on the downwind side of one wood, within touching distance of it, and you will find yourself largely sheltered from the winds. Now walk out into the open area, toward the next set of trees. As you walk, try to gauge the wind speed. There will be some interesting fluctuations in direction as you begin your short journey, but we will return to those. For now, consider only the wind speed. Notice how it is strongest and most consistent near the midpoint of your walk, dropping and then disappearing as you draw closer to touching the trees again. You have walked into a second wind shadow, the one on the windward side of the barrier. It is very rarely noticed or even imagined but is an important part of the weather experience in every landscape.

I spent the rest of the day out in the hills, roaming, thinking about the Rise of Northwood, watching the clouds change and pass. My walk took me onto a road that in turn met a junction with a farm track. The two met at the corner of a field where an open gate led into stubble.

The hedges that ended at the gate funneled the wind: It poured through this opening and pushed me back. A few paces to my left, the wind shifted, weakened, and followed the track. To my right it was weaker again and hugged the road, rolling along the pavement between two high hedges. I walked through the gate and onto the stubble. The wind weakened and shifted yet again.

The wind we feel is molded by our landscape. But it also shapes it, as I had witnessed in the withered saplings of Northwood. All of these winds are driven by the pilot of the upper winds. The higher they are, the more confident we can be that they will faithfully forecast any major changes in the weather. But our experience is richer for noticing the smaller changes we live among.

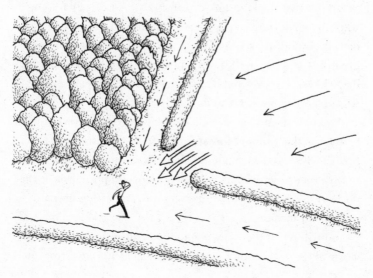

CHAPTER 6

Dew and Frost

DEW

A Stroll into the Light • *Gideon's Fleece* • *A Romantic Myth* •
Moisture and Murder

For many years I had mixed feelings about dew. I'm sure you've had similar emotions. You wake up in a tent after a clear, starry night and sense the rising sun through the fabric. You stretch, and unzip the tent, looking forward to the twin joys of camping—warmth and dryness—only to find that the ground is soaked with dew. It's beautiful for sure, but it's wet. Everything left out on the ground is soaked—well, not quite everything: Some things are dry, which is odd. My ambivalence toward dew changed to delight when I learned how to read it.

We know it and yet we don't. Everyone recognizes dew, but very few expect to find any meaning in it. Scientists struggle to predict where and when an individual dewdrop will form, but it's much easier to understand why dew blankets appear where they do. Within the blankets there are patterns, and within the patterns we find our meaning.

You will recall the "dew point," the temperature at which the vapor in the air becomes saturated and condenses to form water.

There is always vapor in the lower atmosphere, so there is always the potential for dew if the ground is cool enough.

If the air falls to this temperature, water droplets appear in suspension and fog forms. But before that happens the ground will cool more quickly than the air, and any vapor in contact with it condenses, leaving familiar water droplets. Dew is therefore a sign that we are not far from fog, but we will always see dew first. It is possible to have dew and no fog, but not fog without dew.

For ideal dew conditions we want very moist air and cold ground. The ground loses most of its heat during the night under clear skies. This is the first anomaly and clue and it's a seasonal one. We need lots of moisture in the air near the ground, but clear skies are a sign of dry air in the atmosphere. It often happens that the ground is cool, the lowest layers of the atmosphere are moist, and the higher layers are dry, and that's most common in autumn.

Dew is more likely when the ground contains lots of moisture. It is rare in dry regions when there has been no recent rain. In autumn, the summer's dry spells have passed and the ground is growing wetter with seasonal rains. It is still warm enough during the day for the air to carry lots of this water as vapor and there are still enough clear skies for lots of heat to escape at night. We have our perfect dew cocktail: clear skies, moist lower atmosphere, moist soil, warm air cooling fast, and cold ground.

Dew is associated with calm night conditions. The ground needs to cool, and this is much less likely if it is windy. Wind stirs the air near the ground: We need the layer nearest the ground to be allowed to radiate its heat into space and for this cold, very thin layer to remain undisturbed. Wind ferries in warmer air that hasn't been in contact with the cold ground.

Early on a dewy morning I like to hold my hand low, a touch above the dew, and feel the temperature of the air there. Then I raise my hand high above my head: I can often feel the temperature rise in that short distance.

As the sun rises and warms the ground, the dew evaporates, which has a cooling effect. The air above dew-coated land stays cooler for longer than it does in nearby areas without dew. It is sometimes still possible to feel this cooler air near the ground in the middle of the day.

You have, I'm sure, enjoyed the beautiful sight of the morning sun catching the dewdrops on a spiderweb. If we think about this, we will see many of the ideal conditions brought together: a sunny morning, which means continued clear skies, and calm conditions, which allow spiders to spin webs that survive. And the suspended dewdrops are a clear sign that there have been no gusts recently.

If you focus your attention on a dewdrop you may catch bright colors in it. Try walking toward the dewdrop and away again and see if the colors shift. It's possible to see them segue from blue to green, yellow, orange, and red—the same sequence we see if we follow the colors from the inside of a rainbow. The colors in a dewdrop appear for the same reason that we see them in rainbows: The water droplets take in the white light of the sun, then bend and separate the different color wavelengths, sending each out on its own path. When we move slowly toward a dewdrop and see the colors shift, we are crossing these tiny paths of light.

Dew offers us another stunning optical effect. Morning sunshine and dew are common partners, and the next time you see your shadow on a dewy lawn, look carefully at the light around your head. You may well feel blessed. On bright dewy mornings it is very common to see a halo around the shadow of our head, an effect called the heiligenschein (German for halo, pronounced HIGH-lih-ghen-SHINE). With the sun on your back, its light is bouncing into your eyes everywhere you look, but the effect is emphatically strong when you look in the opposite direction to the sun, and this is exactly what we do when we look at the shadow of our head.

A Stroll into the Light

It's worth a short detour here, because we've stumbled on what I believe to be one of nature's lesser-known universal laws. It concerns the sun, which is never far from our thoughts when we scan a scene for weather signs. There is nearly always an unusual brightness to be found when we look directly opposite the sun, and this appears true over extraordinary scales, from dew near our feet to the moon. I first came across this effect when investigating how we can identify a full moon. A moon is full when it is opposite the sun, and the key here is that the moon's brightness shoots up as it reflects so much more light at this time. When full, it is much brighter than it is a day on either side. But between the dew and the moon, this effect crops up in many unexpected places, throwing brightness into every kind of landscape.

A few weeks ago, I was using natural navigation to cross a wild patch in the Scottish Highlands when I spotted something I'd never noticed before. I was on the north side of a small mountain, in the shade of its peak. I looked to the north to try to get a sense of direction from the shadows on the hills before me and was suddenly struck by an unusually bright patch of woodland on the mountain opposite. At first I thought it was a change in tree species, but I was looking at pines with one narrow band appearing much brighter than their neighbors. It took me a minute to appreciate that I was looking in the opposite direction to the sun, which caused the brightness. A few degrees on either side, and the trees lost this luster. I've been experimenting with this since then, and it works when the sun is low enough and its light hits a steep surface, like a wall of trees. It's hard to describe the joy of noticing simple signs that have been hiding under my nose for half a lifetime.

Even the air appears brighter when we look opposite the sun, as it reflects the sun's light back to us. Near sunrise or sunset on a clear day, try looking at the horizon in the direction of your shadow and you will notice brightness in the sky: airlight.

Gideon's Fleece

For the ground to cool enough for dew to form, the heat has to be able to radiate all the way from the ground to the sky. We need skies that are clear of clouds, but there mustn't be anything trapping heat nearer the ground either. Dew flourishes when there's a clear line from the ground to space. It's less likely to form under any cover, even a few tree branches. If you see heavy dew on the grass, follow it to a tree or anything that overhangs the ground and you will find it thins or disappears. The air and ground where the dew stops will be a few degrees warmer: You will feel it if you walk over this line in the still air of early morning.

We see where the dew starts under open skies and stops under any cover, but why does it start and stop so suddenly in places where there is no cover? This poses a new mystery for us to solve. In many gardens it is common to find dew layered generously over every blade of grass on a lawn, only to find it stops at the edge of a flowerbed. The same thing happens in wilder landscapes, too, with dew covering every leaf of the low plants but avoiding the soil and rocks nearby. But why?

The dew is trying to tell us a story about the ground all around us, and we can learn to read it by thinking about Gideon's fleece.

Then Gideon said to God,

"If You are going to save Israel by my hand, as You have said, then behold, I will place a fleece of wool on the threshing floor. If there is dew only on the fleece and all the ground is dry, then I will know that You will deliver Israel by my hand, as You have said." And that is what happened. When Gideon arose the next morning, he squeezed the fleece and wrung out the dew—a bowlful of water. (Judges 6:36-8)

Gideon found his fleece soaked with dew but the ground all around it dry. Was he asking for a sign or did he know his dew and therefore that this was predictable? A discussion for another time, but Gideon's fleece helps us unlock the lower map that dew is making. For dew to form, we know that the ground must cool to below the dew point, but there are always two parts to the cooling. It is not just a question of how quickly heat is being lost to the sky but also of how it is being replaced from below.

At ground level, heat is always being lost upward, but it is also flowing up from underground, too. The amount of heat that reaches the surface from below is determined by how well the ground conducts that heat. Some surfaces, like soil and sand, conduct heat quite well; others, like plant leaves, do not. The heat radiates out to space from all open areas, but some are topped up by the warmth from below and others are not. The soil cools, but it also draws warmth from below, which keeps it just warm enough to keep dew from forming. The blades of grass lose a lot of heat but are not topped up from below, as heat cannot flow well through the plants. The result is that the surface of plant leaves ends up much cooler than the soil nearby. The blades of grass fall below the dew point and gather dew; the soil doesn't.

Gideon's fleece and the land all around lose heat overnight. The ground heat is replaced from below, but the fleece is made of wool, one of Nature's great insulators. We make clothes from wool specifically because they are so bad at conducting heat away from our bodies: They keep it trapped and us warm. In this experiment, the wool keeps the heat from the ground from rising. The dew forms over the cold top layer, soaks in, and is continually replaced until the fleece is sodden.

My sons regularly repeat their own version of this experiment after playing in the garden. They forget the shirts they've taken off in the heat of their games, and the following morning we pick up sodden clothes from the lawn, even when there has been no rain overnight.

A Romantic Myth

A popular myth has it that dew can deposit enough water to fill ponds and quench plants in the way that rain will. But too little water is suspended in the lowest layer of the air for this to hold true. Some specialist organisms, a few desert plants, lichens, and some pines, survive courtesy of dew, but for plants in temperate zones, dew plays a minor role.

In the South Downs I regularly come across "dew ponds." Farmers used to dig hollows and line them with clay, to offer a watering hole for the sheep on these dry chalk hills. Sheep drinking the overnight dews is a romantic notion, but it is a misnomer. Dew can't fill the ponds: only rain can.

Moisture and Murder

A thousand dinars to the creature that can walk across a dewy lawn without leaving a trace! Just try it—impossible! Dew is Nature's way of leaving fingerprint powder all over the scene before the suspect has even ventured onto it.

Some surfaces are great for detailed impressions of animal tracks. Snow can remember where a bird has taken off, and I recently spent a happy quarter of an hour tracking a heron over damp sand, matching the prints it left to its actions. But dew is another beast altogether. Like a sherry-filled aunt at Christmas, dew remembers everything, but is vague with detail. A million tiny droplets and a bright, low sun reveal every time something has touched the ground, but surprisingly little about them. The stories in dew are tantalizing, sometimes a tease, and only occasionally used to solve murders.

One August morning in 1986, police in Pennsylvania received a call from a distressed man. When they arrived at the home of Glen and Betty Wolsieffer, she was already dead, having been beaten to death in the main bedroom. Glen told them there had been an intruder and that he was injured after the intruder had attacked him from behind.

Police found a ladder outside the house, and Glen suggested the unknown intruder had scaled the ladder and gained access through a first-floor window. A few things made the officers suspicious of Glen's story, and two of those things involved dew. It had been a cold, clear night, and dew had accumulated on the roof that the intruder was supposed to have clambered over to reach the window, but there were no footprints in it. Also, police noticed two cars in the driveway, Glen's and Betty's, but only one had a coating of dew.

Police knew Betty's car had been parked in the same spot overnight, as it was covered with dew. Glen claimed he had gotten back from a night out at 2:30 AM and that he had not been out since. But his car had no dew on it. A forensic meteorologist helped police to the conclusion that Glen had been out since the dew had accumulated on his car, which would have been well after 2:30 AM. The warmth of the engine and the air flow had dried the vehicle.

That Glen was in the habit of cheating on his wife, and having more than one affair at the same time, did not help his pleas of innocence. He was found guilty of murder and sent to prison. The true-crime documentary of this story was titled *Dew Process*.

FROST

Rime Ice • Glazed Ice • Hoar Maps • White Contours • The Thaw

Dew has a cold cousin called frost. Most of the frost we see forms in a very similar way to dew, but when the air is cold enough, the water vapor freezes to ice crystals. This type of frost is known as hoar frost.

Hoar frost is dew that has frozen, but the moment it freezes is important. If the water freezes as it is deposited, then a layer of frost will cover a leaf or other surface, leaving a familiar coating of white ice crystals, which are spiky when seen up close. If, however, dew is deposited and the temperature drops until the dewdrops freeze, a slightly different pattern is formed, with ice that is less brilliantly white and sometimes harder to see. It is this slower-forming frost that creates the beautiful ferns that spread over windows and other flat surfaces.

Since hoar frost is formed in the same way as dew, we are most likely to see it in similar conditions, after a night of clear skies and very low winds. It will be found in open areas on surfaces that do not conduct heat well and will disappear under the shelter of any cover, like trees. Walk from under a tree over soil to grass and you will often find the white growing, from no frost under the trees to specks on parts of the earth to a blanket over the grass and a thick coating on thistles.

Rime Ice

A common type of icing is regularly called frost, but is not formed in the same way. Rime ice occurs when very cold airborne water droplets are blown onto very cold surfaces and freeze on contact. Since rime ice forms as a wind blows, it creates telltale asymmetries: It is thicker and heavier on the windward side of any object it freezes on. Sometimes this creates elaborate sculptures of ice, daggers, or feathers that jut out in the direction the wind has come from. It can build up such thick layers that the weight can cause damage, as the naturalist W. P. Hodgkinson noted in 1946:

> The leaves, grass and twigs became encased in cylinders of ice, so that the trees swayed in the slightest breeze and gave out tiny tinkling sounds like so many glass chandeliers. The effect was fantastic and unforgettable. Many of the trees became so top-heavy that the branches fell to the ground under the tremendous burden of ice.

Rime ice is formed by a wind carrying air so cold and wet that it is normally foggy, too. The icy fingers can be used to navigate, as the direction in which they point will be consistent over wide areas. They have helped me on some walks when visibility was poor: From heavy branches, hundreds of cold, white crystal fingers pointed the way.

Glazed Ice

When water freezes on the ground it may form a translucent sheet of ice, which can be very hard to see. It is sometimes referred to as "black ice" on roads, but in this instance the blackness comes from the road, not the ice, which is near transparent. You may also hear it called "glaze" or "glazed ice."

Hoar Maps

Rime ice can be picturesque, and there is real artistry in all frosts, but we will be focusing on hoar frost, as it holds the most interesting signs. The cloudless skies and still air that favor dew and frost are most likely to occur during a high-pressure system, well outside midsummer.

Hoar frost draws a familiar map on the ground. It can form only if heat is not flowing too freely from the earth. I walked around the Scottish town of Inverness one February morning when a harsh hoar frost had seized the land. I followed the river Ness to the edge of town, then ventured out across some playing fields to gain a better view of a pillar of steam that was rising from a factory on the other side.

The steam rose to a glass ceiling, then slid along under it for a couple of hundred yards before dissolving and disappearing. These were classic temperature-inversion conditions. The steam hit the ceiling where warm air lay on cold and spread out. So much heat had radiated out from the ground overnight that the air nearest the ground was now much colder than the air higher up. If you sense these conditions on a cold, clear morning, it is well worth reaching down to feel the air closest to the ground, then stretching up again. There may be eighteen degrees' difference!

It is ironic, but the inversion layer during a sharp frost is so robust that commercial fruit growers sometimes use it like a greenhouse roof. Small heaters are placed between the trees and the heat rises, but only until it hits the inversion layer, which, at the coldest times, may be only a little more than ten yards above the ground. The heat is trapped under this invisible layer, keeping the air near the fruit above freezing.

At the corner of the sidewalk in front of a nursing home, cobblestones were packed among grass and mosses. The plants were white, but the stones conducted heat from below and were dry and clear.

My walk took me past the site of a recent, probably illegal, fire in the middle of some playing fields. It was clear that a jumble of wood had been burned, leaving an assortment of charred timber and blackened metal fastenings. I spent a moment trying to piece together some object from the dark waste and ashes, but my industrial forensic skills failed me. It was time well spent: The riddle stopped me long enough to notice that the frost that blanketed the grass contained many messages.

On the dark bare soil there was no frost: The heat was flowing up too easily from below. The leaves and wood fragments all had a coating of ice on their highest parts, but not near the ground. Even the ash keys—sometimes known as helicopter seeds of ash trees—had their share of frost on top. The metal was mostly clear. The old fire had sharpened my focus and now small patterns emerged wherever I looked. Nearby, one curled brown hazel leaf had frost on its top, but underneath there was a bright green patch of warm-looking frost-free grass. The leaf had played blanket to that tiny bed.

Once we have tuned in to them, patterns that were invisible yesterday shine out today. Of course, no sooner do we start to see them everywhere than we realize we are joining those who opened their eyes before us. We join the Gang of the Observant, many of whom are now long departed. Gilbert White wrote in his classic *The Natural History of Selborne*:

> Old people have assured me, that on a winter's morning they have discovered these trees in the bogs, by the hoar frost, which lay longer over the space where they were concealed, than on the surrounding morass. Nor does this seem to be a fanciful notion, but consistent with true philosophy. Dr Hales saith, "That the warmth of the earth, at some depth under ground, has an influence in promoting a thaw, as well as the change of the weather from a freezing to a thawing state, is manifest, from this observation, viz. Nov. 29, 1731, a little snow having fallen in the night, it was, by eleven the next morning, mostly melted away on the surface of the earth, except in several places in Bushy-park, where there were drains dug and covered with earth, on which the snow continued to lie, whether those drains were full of water or dry; as also where elm-pipes lay under ground: A plain proof this, that those drains intercepted the warmth of the earth from ascending from greater depths below them, for the snow lay where the drain had more than four feet depth of earth over it. It continued also to lie on thatch, tiles, and the tops of walls." Might not such observations be reduced to domestic use, by promoting the discovery of old obliterated drains and wells about houses; and in Roman stations and camps lead to the finding of pavements, baths and graves, and other hidden relics of curious antiquity?

There is nothing new under the sun and few who take the time to see any of it.

Every terrain has its own frost profile. Grasslands will be much colder and frostier than fields, but heaths, dry reed beds, and drained peat bogs are notorious for super-low overnight temperatures. A drop of more than 9°F (5°C) is common over short distances.

If you pause to look for it you will also notice how sensitive frost is to height. It is easy to spot: Any parts of a plant that are more than a couple of feet off the ground are often frost free, while the lowest basal leaves have a generous white coat. Between these we find that angles are important. Notice how tall blades of grass have frost on their horizontal but not their vertical blades—flat blades radiate heat upward more effectively. As the sun comes up, the frosted blades that stand a little proud catch the first rays and are thawed, the ice crystals turning to dew. The lower blades hold on to their frost until the sun is higher. You'll know you've tuned to the finer frost map when you can't help but spot the "frost bounce": Blades of grass and other leaves that were locked in position by the ice bounce up as they thaw in the warming light.

White Contours

On a small scale, dew and hoar frost favor the same areas, but if we step back and widen our focus a little, we discover that the frost has a game all its own. In areas that see plenty of dew in autumn, some parts will be prone to hard frosts in winter and others very nearby never see it. These neighboring areas might share the same open sky, even the same soil and plants, but even so, there are huge differences in the frost map. Why? The answer lies in the shape of the land.

Land is never perfectly flat, and its shape makes a huge difference to frost. Convex landscapes, like hills, are frost-unfriendly, and concave landscapes, like valleys, welcome it. There are two main reasons for this. The first is that hills are exposed to the wind, which dries the ground and mixes the warmer, higher air with the very cold air below, making frost less likely. But the main reason is that cold air is denser than warm air and will flow downhill, like molasses.

Let us imagine a hill and a neighboring valley during a very cold starry night. The heat radiates out of both in similar ways. We now find that we have a layer of very cold air just above the hilltop and the valley. Gravity pulls the cold dense air downhill. The cold air in the valley has nowhere to go and sits there, but the cold air on the hilltop will flow down toward the valley. This has two effects: It leaves the hilltop with warmer air and adds another layer of cold air to the valley, sandwiching that in. This is why some areas are prone to severe frosts and are known as frost pockets or hollows. It is also the reason that hilltops are often a lot warmer than valley bottoms at the start of the day after clear skies. Amazingly, the bottom of a valley can be colder than the air 10,000 feet (3,050 m) above it, which should give food for thought to those in survival situations.

When our lives are not threatened, we can take our time and enjoy the perverse pleasure of feeling temperatures rise as we walk up hills after a cold clear night.

We learn lots about frost by looking to the extremes. Let's start with places that frost avoids altogether. Small islands and coastal land rarely experience it, as the sea warms the air too much. The same effect on a smaller scale can be seen as a frost-free strip near rivers. During my Inverness walk, I found a couple of yards of dewy green grass nearest the river, contrasting with the white sheen of the land all around.

Every region will also have its extreme frost pockets where record low temperatures are noted. Wherever you are in the world, they are likely to share certain features. They are inland and at valley bottoms. There is a famous one in the Chilterns, a modest range of hills lying northwest of London, about an hour away. Their highest peak is well under a thousand feet, so it is hard to imagine their claiming any extremes. But unlikely extremes teach us lots, too.

The Chilterns' frost hollow is at Rickmansworth. It is well inland, and the pocket is in a valley, although not the lowest spot in the region. The local quirk is that as the cold air flows down from the hill into the dip, it is stopped in its journey by a railway embankment on the other side of the hollow. If we return to thinking of cold air behaving like molasses, it is easier to visualize how it is trapped by a small barrier.

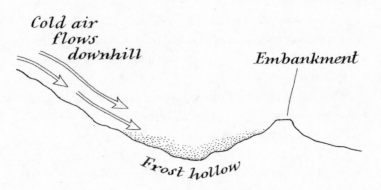

The effect works on much smaller scales. This morning a frost coated the land, and I paused to study it on my way to the woodland cabin where I was writing this book. There was more frost in the gentle dip in the farmer's fields than on the higher parts, but near my feet there were subtler patterns. The frost was thick on a broad thistle, and I enjoyed noticing how this prickly plant had played the part of Gideon's fleece. The leaves just off the ground had kept the earth's heat at bay. There were only tiny flecks of frost in the soil nearby.

There is a gentle roll in our small lawn, and the frost had avoided the higher parts altogether, preferring the lower grass at the border. The lawn is about 65 feet (20 m) across, and the change in height can't be more than 3 feet (1 m), but this is enough for the frost to express preferences. There is paving between the grass and the house, which isn't perfectly level either. Rain from earlier had gathered in a shallow indentation in one of the stone slabs and had frozen overnight. It was so perfectly clear that at first I took it for liquid water, a shallow puddle. It was only when the toe of my boot slipped as I tested it that I appreciated that it was solid and transparent, black or glazed ice.

I forgive you if you have reservations about whether we really will see changes over such a small scale and will not be convinced until you witness them for yourself. But if we look at a tiny controlled example, it may nudge you toward a greater faith. Scientific experiments have shown that a polystyrene box only 3 feet by 3 feet (1 m by 1 m) creates a shockingly different microclimate. Over ten nights, the temperature in the box was found to be about 12°F (or nearly 7°C) lower on average than the air outside it. The box acts as a miniature valley and insulator against ground heat. In seasonal terms, climbing into the box at dawn would be like rolling from June to November in a few feet. We can see why meteorologists stipulate that instruments must not be placed in hollows. It makes perfect sense but is part of a weather culture that hides the rich diversities surrounding us.

Frost can have a devastating impact on plants, so farmers quickly learn its habits. In some regions frost is contentious. The indigenous Wola people of Papua New Guinea occasionally find themselves in conflict with neighboring group, which has led to a culture of fierce rivalry. They will celebrate the misfortune of their rivals, and this includes singing when an enemy group is forced to live off land that is known to be a frost pocket. The Wola habit of rejoicing in frosty schadenfreude has its own name, *liywakay*. It seems a long way from our culture—or maybe not. I can imagine a prize-vegetable grower in a village in the UK dancing gleefully on hearing the news that their bitter rival has lost a crop to Jack Frost.

Frost pockets can explain why some plants at the top of small hills come into leaf earlier than those lower down. It seems counterintuitive until we realize that their height advantage means they are spared the late frosts. In severe pockets, shoots can be killed off altogether. We can spare ourselves the bitterest temperatures when camping on cold, starry nights by avoiding sleeping in hollows and moving a few feet to higher ground.

The Thaw

Frost and brightness go together. The clear night skies that cooled the land mean the sun will soon be lighting the white, which gives us a new range of patterns to look for, many predictable and dependable, a few less so. We find frost lingering in shade, of course, so as the sun fills much of the land with color, the west sides of hills, trees, and buildings will hold their frost longer in the morning, and the north sides may keep it all day.

Every surface reflects the sun's warming rays in different ways and amounts. Frost reflects light well, which is, of course, why it creates such striking sunrise landscapes. But darker patches warm more quickly and thaw the frost nearby: The patterns we see will change by the minute.

Without analyzing every possible combination, I'd encourage you to notice where frost disappears first and lingers longest, and to study this over different scales. It clings stubbornly to the bottom of valleys, but a bit higher up it is found on the ground to the north side of trees and rocks late in the afternoon. Zooming in further to where it only just survives in tiny patches, we find that every tiny frost island is telling us something about the altitude, the sun's journey and therefore direction, and the wind, as well as the contours and colors of the land in that one spot.

Rain

RAIN IS NOT THE HERO of many books, but neither is it the villain of this one. When we know what to look for, rain can tickle our curiosity as well as our necks. We can search for lots of small, subtle characters in rain, and two big, bold ones. We will meet them all, once our senses are ready.

The raindrops we see are the end of a longer journey: A lot of rain starts life high above us as snow or ice. The water picks up particles on its way down, and these give it a signature flavor—not all rain tastes the same. Theophrastus, the ancient Greek philosopher, noted how coastal rain in Greece tasted salty, especially if it came with a wind from the south. Sadly, a study in the Rocky Mountains in Colorado found that 90 percent of rainwater samples contained microplastics—the tiny pollutants were found even high in the mountains. After a long dry spell, rain acquires a familiar, unique odor, "petrichor," as it mixes with oils on the ground and bacteria in the soil.

When rain hits the ground, the nature of the terrain changes the sounds it makes. Rain landing on mosses sounds different from rain falling on rocks, and both are much quieter than any drops landing in a puddle. Soft rocks, like chalk, sound softer than harder rocks, like granite. Gradients shape the sound of rain, too: the steeper the slope, the softer the sound, until rivulets form and turn up the volume. The sound evolves as the rain continues. Listen carefully and you can hear the sound of dry ground changing over a couple of minutes under heavy rain.

Each species of tree sounds different under rain, and this also changes with time. Deciduous trees sound very different across the seasons, of course, but also at the start and end of any shower. Conifers are quieter at the start of any rain, whereas broader leaves give out more sound.

My favorite rain-sound maps are found near isolated trees. I like to stand under a broad, thick conifer, perhaps a yew, and listen to the rain as it starts. The tree offers a rain-and-sound umbrella, which makes listening to the patter of drops on the ground nearby much easier—the tree shields our ears from the sounds of rain on our shoulders or clothing. On the forest floor near my home a mix of brown, dead leaves forms a carpet that is broken by the green bracken, brambles, and ivy leaves. Dead, dry leaves make the most noise when hit by a large raindrop, emitting a percussive rap. Bracken whispers weakly.

The gradient of leaves shapes sound just as the gradient of land does. Bramble leaves point downward and hang with a softness that is reflected in the faint sounds they make—the weakest noise you can make by blowing air to part your lips. The ivy leaves that carpet the woodland floor are closer to horizontal and have a firmness in texture and support that seems to place them between the dead beech leaves and the brambles. Raindrops hitting dry dead beech leaves make a sound like a pencil point rocking from a hand onto a well-read newspaper (the pages must be loose, not iron fresh). Forest-floor ivy leaves under rain are fingertips falling onto a glossy book cover.

My habit is to find a patch with a mixture—the exact ingredients change each time—and listen with eyes closed, then open, until my ears understand the same things as my eyes. Then I walk on until I find another sound-umbrella conifer and repeat the exercise. This time, though, I close my eyes quickly and let my ears tell them what they will find when I reopen them.

RAINBIRDS

When we walk through woods after rain there will always be secondary showers, as some of the rain that is being held in the canopy is shaken loose by the wind. We can hear the breeze responsible, as it shakes the tops of the trees. But there are other even gentler secondary showers that sound and feel different, and they encourage us to look up.

When the rain stops, the water on the upper branches accumulates in a precarious equilibrium. The leaves hold the perfect amount of water for that moment, some drops sitting on the leaf surfaces, others hanging from the tips. The water will sit there until it dries or is disturbed, and because the balance is finely tuned, it doesn't take much of a disturbance to shake it free.

I first started to notice the showers caused by birds taking off when I paired the loud flapping of wood pigeons with the fat dollops of rain that fell onto my head. But since I spotted the racket and rainfall of these rambunctious birds, I have learned to sense the odd lighter shower that is too local, too narrow, too delicate for a breeze. Looking up, I have seen woodpeckers, corvids, and even small songbirds landing or taking off.

SHAPES, PATTERNS, AND TIME

Look at the tips of the broad leaves on the trees you pass. Did you know that the more pointed the leaf tips are, the rainier an area is? Leaves with distinct points have evolved to channel off the rain water more efficiently via a central rib that leads to this point. Tropical rainforests are packed with pointy leaves.

Rain stipples the softest ground, leaving familiar pockmarks in mud, sand, silt, or snow. They tell us about the character of the rain: hard or soft, short or long. The bigger the gaps, the shorter the rain. In soft mud or sand, try to notice the difference between regular, lighter prints of rain and the deeper, less regular marks of secondary showers. These imperfect, cruder raindrop prints show us where a breeze or bird has shaken down the heavy drops. An hour ago, a crow took off from a branch fifty feet above our path, yet that story, written in rain, is still fresh by our feet.

Raindrop prints can be a time stamp in other stories. If we remain sensitive to the recent history of rain, we can use it as a clocking-in card for animals or humans. The rain's footprints lie over those of any animal that passed before the shower and are obliterated by any that came after it.

It is easy to lose sight of animal tracks and think the game is up, only to remember that you're under a dense tree canopy. With intelligent guesswork, we can look for the place where the animal might have emerged from the umbrellas and find their prints reappear in the softer, stippled mud at the edge of the trees.

Many of rain's stories are fleeting and ephemeral, but one allowed me to revel in a giddy sense of time travel. At the Shrewsbury Museum, in Shropshire, there are small fossils in stone, rock-hard pockmarks that reveal how rain fell on that spot more than two hundred million years ago. Marks I had associated with the past few minutes told a story from many millions of years ago. It gave my mind a healthy spin.

RAGGED BOTTOMS

Is it going to rain? What a popular question. The key to longer-term forecasts lies in the earlier chapters about clouds and fronts, but we will often find ourselves looking at an individual cloud and wondering if it's about to soak us.

Rain rarely falls from small cumulus clouds, and if they're wider than they are tall, it's very unlikely indeed. But we can be a bit more forensic about this by studying the bottom of the clouds. As we have seen, the bottom of all clouds marks the height at which the temperature is cool enough to condense the water vapor in the atmosphere, the dew point. Since this level tends to be consistent, it leads to flat-bottomed clouds. It follows that a cloud with a ragged bottom is trying to tell us something.

We're all used to the idea that clouds give us rain. It is less well known that rain creates clouds. When rain falls, it cools the air just below the cloud. This leads to more condensation and creates jagged areas of new cloud just below the main cloud, which gives the base a ruffled, rough appearance. Hence, smooth bases on cumulus clouds mean no rain falling, and ragged bottoms indicate rain falling. These craggy, uneven fragments of cloud are known in formal circles as "pannus" and are described as "accessory clouds." They are rain's footprints and can be seen under all clouds that rain falls from.

However somber a cloud appears, if it has a neat horizontal base and good visibility below, it is unlikely to rain on you.

RAIN GHOSTS

From time to time you may see what look like thin streaks falling from the base of clouds: virga. This is rain that evaporates before it reaches the ground—rain we can sometimes see but never feel. Virga consists of droplets or ice crystals that are just large enough to fall but not substantial enough to make the journey through the drier air below to the ground. If there is a strong wind at cloud level, the streaks can be seen falling behind the cloud as they drop into the slower wind below them. Although most common in hot, arid regions, virga can be seen trailing from the base of clouds anywhere.

Virga is an in-between sign: Conditions are nearly right for rain, but there isn't enough water in the air yet. Like so many weather signs, virga is most useful in noticing a trend. Virga after heavy rain is common and is part of an improving situation; after clear skies, it means rain is not far off.

The way the rain trails behind the parent cloud reminds me of cartoon ghosts that float above the ground, trailing their lower bodies behind them. This, combined with the way virga clouds are both raining and not raining, has led me to think of them as rain ghosts.

BLANKETS OR SHOWERS

Drizzle, mizzle, Scotch mist: There are lots of words for rain, but they mostly refer to nuances in the size of the raindrops or the visibility. Drizzle is defined by some as raindrops smaller than 0.51 millimeters and can be categorized as light, moderate, or heavy, depending on visibility.

These definitions can be useful if you're writing technical reports or love to play with arcane vocabulary, but they don't change what we see or what it means. As sign readers, we want to avoid getting bogged down in names and to keep things simple. We'll stay focused on the key fact: There are only two types of rain, blankets and showers. Neither word sounds precise, but there is an important distinction here: Blankets of rain are wide and last a relatively long time. Showers are short.

Neither word refers to how heavy the rain is. "Shower" is used colloquially to refer to light rain, but in fact we mean something quite precise. Showers can be very heavy downpours or the lightest of dots on a windshield, but in all cases, they're short-lived. You can't have a shower that lasts an hour.

Why is the difference between blankets and showers so important? You will recall that there are blanket clouds (the stratus family) and heaped convective clouds (the cumulus family). And you will not be surprised to learn that those two cloud families lie behind the two types of rain we experience. Once we have identified the type and its parent clouds, we will know so much about the formation and history of that rain that it instantly allows us to deduce and predict a lot more.

Just the simple division of rain into blankets and showers is like getting to know and recognize two different species. It's a fair analogy. If someone is unfamiliar with deer in my local woods, they might see a brown quadruped bound away into the distance and say, "Look, a deer. I wonder if we'll see any more." But knowing that in these woods you will see either roe or fallow deer and learning a couple of simple tricks to identify them allows us to say, "Look, a fallow. Where are the others? They must be out there somewhere." Once we know that fallow are gregarious social deer and rarely found alone, the sight of one leads quickly to the prediction of others. The same approach works with rain clouds.

As soon as we see, feel, suspect, or know that rain is imminent, we just need to decide if it is a blanket or a shower. If the former, it will rain for a long time, but with a light to medium intensity. If the latter, it will be short-lived and could be anything from light to very heavy. If we see stratus clouds, we will lean toward blanket rain; if we see heaped clouds and patches of blue sky, showers are the likely suspects. In both cases, the depth and color of the clouds will flesh out the picture.

If you suspect that showers are occurring in the area, check the cumulus clouds regularly. Their numbers, shape, and size will change quickly, but as a general rule, once there are more heaped clouds than blue sky, showers are likely.

When it comes to stratus clouds, the rain-bearing layer will arrive, get steadily lower, then block your view of most other clouds. In this case you can look for color and texture differences in the layer, but you'll have to stay tuned to wind direction and an understanding of fronts to call the broader patterns and progression.

The other day I was out in the hills with a small film crew. We were building an online course to help me teach during the COVID-19 lockdown. I had been keeping an eye on the towering cumulus clouds all day and knew what the afternoon had in store. It was no surprise when a heavy downpour joined us near the top of a hill. The dry puddle basins began to fill, and the sounds changed by the second.

The cameraman was understandably worried about his equipment, and we did our best to shield it with an umbrella, but the rain was now fierce and coupled with gusts of wind that whipped the water around the shield. The umbrella buckled and the cameraman grimaced. Whenever we find ourselves under a heavy shower, it's a good idea to remind ourselves of how local they are.

Against everyone's instinct, I asked them to follow me along the path, and we walked out of the shower less than a minute later. From sunshine, we were able to watch it soaking the land we had been standing on. The line where the shower started was so clear that I was sure I could have held one hand in the rain and the other in the sun.

RELIEF RAIN

We are now comfortable labeling our rain as either blanket or showers, and happy that this will give us a flavor of the length and intensity of any rain to come. The next major piece of the jigsaw puzzle for us to look at is how this fits with our landscape.

The single biggest landscape influence on rainfall is the shape of the land. When humid air is pushed up over higher ground, it expands and cools, water vapor condenses, and droplets form. When this process leads to rain it is known as "relief rain."

Air near saturation needs only a small bump in the land to form clouds; much drier air may need a real mountain. But since all air contains some water vapor, there is always the possibility that high ground can create rain-bearing clouds.

The physics is the same wherever we are in the world, and you'll see similar patterns across the globe. The summit and the windward side of any substantial hills receive more rain than the downwind side. In some parts of the Scottish Highlands, the windward west sides receive six times as much rain as the eastern lee sides.

Yesterday afternoon I spent a pleasant hour watching small cumulus clouds bulging and swelling, as the wind lifted them over the Sussex hills, then dissipating back on the lee side. The clouds grew until they were past the highest ground, then shrank again. Over several minutes I could watch their shapes change. On this occasion, the air wasn't humid enough or the ground high enough for it to start raining, but we were not far off. Another five hundred feet of hill or slightly moister air might have done the trick.

TECHNOCHEATING

There is a sneaky way of using technology to help fine-tune our understanding of relief rain. In many parts of the world it's quite easy to find weather radar online. The websites show a dynamic map of many weather phenomena, including rain, that update every five minutes or so. Rain is good at reflecting back radar energy and shows up on the maps as different colors, according to the intensity of the rainfall.

One weather radar system I use shows blue pixels where there is very light rain, less than 0.5 millimeter per hour; yellows and oranges for 2 to 8 millimeters per hour; up to magenta for more than 16 to 32 millimeters per hour. As rain-bearing clouds pass over hill ranges, you can spot the change in color near the summit ridges, from light to heavy, then back to much lighter. There is a vicarious pleasure in learning about rain when you're still dry and warm. As I write this, no rain is landing on my cabin, but I've just watched part of the Mendip Hills in Somerset get a good soaking. The windward southwest side was orange in the radar image, but the northeast side had only a few speckles of blue.

If you get into the habit of doing this a few times over a period of months, you'll start to notice how sometimes the effect of high ground on these rainfall maps is dominant and at other times barely perceptible. Obviously, big mountains have more effect than gentle hills, but we can find it anywhere when the air is moist enough.

THE TWO HILL RAINS

We are ready to bring the main pieces together, which takes us back to the two types of rain: blanket or showers. Stratus clouds and the rain they bring are shaped by broad atmospheric conditions and fronts spread over very wide areas, often hundreds of miles. Cumulus clouds are a local phenomenon, shaped by differences over small distances and reflecting changes in landscape over hundreds of yards.

Blanket rain moves over the land in a giant sweep. The effect of a few hundred feet of high ground on these enormous systems is modest. Yes, there will be more rain on the windward side, but it will often be marginal, a little heavier and perhaps a little longer lasting, but not wildly different.

However, if we are expecting showers, the effect of high ground is likely to be much more pronounced and critical. Cumulus clouds are always telling us that the air is unstable. The atmosphere is volatile, just waiting for something to set things off. Earlier we looked at the way local heating differences can trigger towering clouds, and air moving over high ground does the trick, too.

As tall clouds approach a hill, they suddenly start growing even taller. Heavy rain showers or even thunderstorms over the high ground are now much more likely. All landscapes have a noticeable effect on showers, but only grand landscapes, like mountain ranges, have a major effect on blanket rain.

If you can find a place where you can regularly watch rain clouds pass over the same hill range, you will quickly come to appreciate the difference between showers and blanket rain over that high ground. There is nearly always an observable difference in shower clouds near the hill, but it's much harder to spot differences in the blanket clouds.

With practice, you may notice color changes in stratus over gentle hills, and even texture and shape changes in the clouds. But long before you're enjoying that, you'll have spotted how shower clouds grow in number and size on the windward side and summit of even modest hills. Look closely, and you'll see that their bases are darkest near the summit, too, especially if it is wooded. This is the moment to look for the signs beneath the clouds: ragged bases, followed by pannus clouds, virga, then rain.

SHADOWS AND FOEHN WINDS

When the windward side of mountains receives lots more rain than the lee, there are side effects. The contrast creates a "rain shadow," an area on the lee of the mountain that receives so much less rain than the windward side that it affects the nature of the region. Many mountain ranges have rain shadows on their eastern side, including the Pennines in the UK, the Alps in Europe, and stretches of the Rockies and the Sierra Nevada in North America.

We're dealing with rain, but one wind is so intertwined with relief rain that we'll meet it here. After rain falls on the mountain, the air sinks on the downwind side, but it is now much drier. It is also compressed by the higher air pressure as it drops, making it warmer. This warm, dry air spills down over the low ground and creates the rain shadow. The warm, dry leeside air is called a "foehn wind." In parts of the world where it is very pronounced it will earn its own local nickname, too. There is the chinook, or "snow eater," east of the Rockies, the ghibli in Libya, and the zonda in the Andes. The helm wind in Cumbria is a foehn, England's only locally named wind.

The air in foehn winds is so much drier than the air on the windward side of the hill that it can dissolve clouds, even on an otherwise cloudy day. This leads to a blue gap in the grey skies on the downwind side of the summit, known as the "foehn gap."

If the conditions seem perfect for relief rain showers and you don't see any, look for any big hills upwind of you. If you find some, then warm, dry foehn winds may be descending on your area, making showers very unlikely because you're in the "gap."

The foehn effect sounds mild, but it can leave its mark. It is responsible for creating the conditions behind many of the wildfires on the European continent and lies behind the wildest temperature swing ever recorded over twenty-four hours in the US. On January 15, 1972, the temperature in Montana leaped from −55°F to +48°F (−48°C to +9°C)!

LAND AND SEA

The two main triggers for local rain-bearing clouds are the sun heating some areas more than others and the ground forcing moist air to rise. Land always warms more quickly and is higher than the sea. This leads us to one of the biggest and simplest signs: More rain falls over land than over the sea.

Showers over the Mediterranean are quite rare, but they are much more common over the lands that border it. On our local beach in Sussex, we regularly watch yachts in sunshine as showers fall on the Isle of Wight, Hayling Island, and our barbecue.

SEASONAL THOUGHTS

There is a seasonal aspect to rain. Yes, most places get a little more in winter than summer, but it surprises some to learn that we have more showers in summer than in winter because of the stronger solar heating, but also because the air is colder in winter—cold air holds less water vapor. Winter-shower clouds are shallower, with lower water content, than the deep, towering summer-shower clouds.

In many places, the differences between showers and blanket rain lie behind the seasonal character of the rain. It means we have to be careful when drawing conclusions about the "raininess" of an area from monthly figures. If Place A receives half as many inches of rain as Place B in August, we may think it's a better choice for a vacation spot. But what if Place B has half the number of rainy days? It can take hours of drizzle for the same volume of rain to fall as a five-minute downpour produces. Some high plateaus, wooded hills, or cities near the coast are notorious for a seemingly never-ending drizzle, but the total rain figures may paint a misleading picture. Miami receives 50 percent more rain than Seattle. It also gets almost 50 percent more hours of sunshine.

The tropics are notorious for this contrast, and it is well known in the travel industry that the edge of the rainy season is a great deal for many travelers. If you're happy to see an hour or two of wet weather in a day, with hot sunshine for the rest, then there are always bargains to be had. The rain statistics may be dire but the experience entirely different.

THE ANATOMY OF RAINDROPS

In stratus and cumulus clouds, the heavier the rain, the deeper the cloud must be. Rain forms in different ways within these cloud types. In stratus clouds, a "warm rain" process is at work as millions of tiny water droplets collide with each other to form decent raindrops—that's why it takes deep clouds to form large drops. In taller cumulus clouds, like the cumulonimbus, the droplets freeze and "cold rain" forms: Lots of water droplets rapidly condense around the ice particles.

The heaviest rain is found in cumulonimbus clouds. They are tall enough to reach a level where water will freeze; once ice forms high up, large raindrops form around the core ice particles, known as condensation nuclei. That is why the biggest, heaviest raindrops feel surprisingly cold—they were very recently ice. Without tall, freezing clouds, very heavy rain is unlikely outside the tropics.

Wherever you are in the world, raindrops have a maximum size. Using a mind-bogglingly complex formula that we can live without, scientists have proved that once the radius of raindrops grows larger than about 6.2 millimeters, they break up. My less scientific definition of the range of raindrop size grows from "forehead cooling" to "runs off nose."

Heavy rain really does hit us with more power. A law in physics dictates that as the weight-to-friction ratio increases, so will terminal velocity. In English, if you drop a marble and a foam ball of the same size off a building, the marble will keep accelerating to a faster speed than the foam ball. The larger a raindrop, the better its ratio and the faster it can travel: A raindrop that is twice the size of another will also hit you much more quickly. You feel like you've been hit by something bigger and heavier, which you have, but the effect is compounded because it is also traveling faster. Bigger raindrops pack more of a punch in every way.

TOWN, WOODS, AND CLIFF SHOWERS: TOWARD A FINER ART

Once you've enjoyed spotting the two broad types of rain and their relationship to high ground, you'll be ready to look for some of the more niche showers out there. More showers develop over or just downwind of towns and woodland than over open rural land.

If the air is moist and unstable, the local differences in the sun's heating of towns and woods can be enough to trigger shower clouds to form. If there is little wind, the rain will fall over the city or woods, but a little breeze will push it just downwind of them.

Urban showers are both similar to and different from the woodland variety. The sun warms the asphalt of the cities more than the surrounding countryside, but this effect is then compounded by the heat life generates. It creates an "urban heat island," raising the temperature of the city above that of the surrounding area and by more than solar heating alone would manage. This is something we will return to in chapter 17, The City.

Aspect can play a role in both urban and woodland showers. Since showers need the warm, rising air of convection, triggered by the sun, they are more likely in the afternoon, after the sun has had time to warm the land. It follows that, in hilly areas, they are more likely over land with a south-to-west aspect, as these slopes are in full sun during the afternoon. Over both woodland and urban settings, old clouds wither and disappear, and new clouds are born in the same place.

Showers sometimes form above cliffs. A cliff face doesn't lift a wind gently, as a smooth hill face does: It trips it up—the wind is sent into a spinning vertical eddy. At the top of this cycle, it can lead to clouds and showers. The wind is sometimes funneled, then forced upward through gaps in the cliffs or other landforms, like stacks and arches. Each coastline has its signature shape and creates clouds and showers in certain spots.

The sun has heated the tarmac at an airport, creating a strong thermal and cumulus cloud. The low base of the clouds indicates humid air, and the shape of cloud reveals how unstable it is. A shower followed soon after.

Using the techniques above, it is possible to look at a landscape you've never set eyes on before and make useful predictions about rain showers. You'll be right some of the time, but enough variables at play can skew things, some of which, like humidity, we cannot sense perfectly. You'll help yourself reach the next level of prediction by spending time in the same landscapes and noting how things play out.

If you're keen on nurturing this art, you'll want to find a spot with some elevation that you can get to regularly. The better your view in all directions, the more fruitful this exercise. Halnaker Windmill allows me good views, up to 12 miles (20 km) in most directions, over hills, towns, woods, islands, and coastline. I have learned more about cloud and shower formation from this one spot than from many others combined.

Take into account the shape and character of the land, the wind, temperature, and feel of the air, and make your prediction. Then enjoy seeing what happens. You will find there are places where showers will develop every time a really muggy breeze comes in from a certain quarter. Then there are much rarer and more fickle spots. You will find there are telltale showers: If they develop in certain places, they're pretty much guaranteed to develop in other dependable spots, too.

CHAPTER 8

A Bloodhound in the Woods

The House of Wild Truth • The Canopy Breeze • Bloodhounds •
The Six Shapes of Smoke

I MET EXPERT TRACKER JOHN RHYDER at the bottom of a hill. John had reached out because he wanted to compare notes about wind behavior in woodlands. He wanted to solve a riddle.

We walked into a hilly, wooded area of the South Downs that he uses for his tracking courses, and he explained how he knows that the fallow deer will be lying on one of two ridges. He knew the areas on the ridges the deer favored, but he was trying to determine the logic for the spots they were choosing as their bed each time, and he felt sure that the wind played a part.

Dull grey stratus blanketed the sky, and the winds were light. Woodlands under grey skies and light winds are problematic for natural navigation anywhere in the world and force our attention toward some of the finer clues. I paused to get my bearings. There was a good compass in the color of the trees. We stood in a grove of young beech trees, only about fifty years old, which allowed enough light in to create a contrast among the colors of the trees when looking north and south.

The trees appear greener when looking south and "cleaner" when looking toward the north. The reason is that the north side of the trees remains a little damper in the middle of the day, shielded from the southern sun by their trunks, which allows the algae to thrive. The algae gives the north sides a green tinge, while the south sides are cleaner, with a sprinkling of pale lichens.*

As soon as I had a compass I could trust, I found a break in the canopy and looked up at the clouds. A uniform grey blanket is only uniform until we give it our proper attention. There is always a little variety, a shift in hue, a rough line between two greys or a lower portion that is a different color from that of the cloud just above it. And these patterns are enough to give us a clue to the direction in which the clouds are moving.

I noticed a lot of jagged edges, a sign that the bottoms of the clouds were broken, uneven, and that some rain was likely. The light grey colors in all directions suggested that any rain would be light drizzle. I tried to gauge the height of the clouds, to give me a chance to monitor whether there was any trend in this. It was difficult in the woods, but before we ventured in too deep, I saw that the base could just be made out above the gentle curves of the hills in the distance.

A forestry track led us deeper into the trees, and we clambered over a storm-felled trunk that blocked our way. The tree had come down in a recent gale and was a mighty compass. Like most fallen trees in the area, it pointed from the southwest toward the northeast, because that is the direction in which most storm winds blow in England. The track curved and forked, and I glanced back regularly. It's always a good idea to look back as paths bend: It gives us a second perspective on a route and

---

* A little confusingly, we see the north side of trees when looking south, and vice versa.

helps us organize things in our minds. I don't know the science that might explain this, or even if it exists, but I can say that this habit seems to allow a totally different part of the brain to register what is going on. Maybe the path ahead, the future, and the path behind, the past, really are separated by the brain. In life's journey, the wise like to tell us that only the present and future matter. But in natural navigation, the future is rosier if we don't let go of our past.

THE HOUSE OF WILD TRUTH

We abandoned the track and made our way up a steep section of hill. Soon we encountered what I, with a smile, like to call the House of Wild Truth. My eyes were pulled in many directions but settled on reading branch shapes. I stared keenly at the contrasting way in which the upper branches of the beech trees were curving down toward the south, while the lower ones were not. I was enjoying a delightful demonstration of the way that land shapes light. We were on the north side of steep high ground, among tall trees. Their tops were open to the southern light, and these branches reflected that in their shape, trending down toward the horizontal on the south side of the tree. But lower down on the same tree, the branches were shaded and did not reveal this trend.

All the time my eyes were jumping from one high branch to another, like an orangutan, John's gaze had been fixed closer to our feet. He pointed to a spot where a badger had cleared the leaves and rooted around in the mud for food. And near this brash sign there were cuter ones. Leaning down, John pointed to a place where the color of the forest floor changed ever so slightly. The light brown of the dead beech leaves turned darker in a spot that was no bigger than a plum. It was a deer track, and near it, others showed the path the deer had followed downhill.

Like so many signs, it was now obvious, but I would have walked past it for sure, my head between the high branches and the clouds. Deer tracks in soft mud call out to me and require no extra focus to spot, but John showed me something I had not tuned in to before. A heavy animal with narrow feet, like a deer, will bend leaves upward as they step on them. It leaves a trail of leaves pointing toward the sky, an anomalous arrangement that gravity on its own would never achieve. A falling leaf will never come to rest on its end. As with so many of these clues, the logic was simple and infallible, and the sign now shone brightly from among the leaf litter around me. The deer might as well have walked through wet pink paint.

For a few minutes, John and I had each focused on our favored signs—equally valid but very different. This has happened to me so many times when walking with experts in other disciplines and it leads to heady mixed emotions. What joy—there's so much more for us to sense and find meaning in! What terror—how much I must have been missing!

To accommodate the two, I've built a thought experiment called the House of Wild Truth. The wisdom and truth of the wild is laid out on an enormous table in a grand mansion. We are not allowed inside the building. Each time we are outside, we're trying to see what lies on the table. We creep toward the house and find a window to peer through. The table is quite far away, but we can just make out what lies on this side. We step back and creep around to look through the next window. We see some of the same things we saw before from a different angle, and a few new things, too. There is no window that gives us a perfect view of the table, but we learn something from each perspective, and slowly we come to understand how the pieces on the table are related to each other. It is a giant three-dimensional jigsaw puzzle, and if we're lucky, we gradually see enough pieces and how they fit together to gain some idea of the full picture.

The light, the wind, the clouds, the shape of the trees, the angle of the leaves, the behavior of the badger and the deer . . . and this was only what we saw through a couple of the windows.

THE CANOPY BREEZE

At the bottom of the hill, in the lee of the high ground, there had been a moment of calm when we had not picked up the faintest trace of a breeze. The air was not still: We were just unable to sense the breeze. If we had been more forensic, we would have mimicked the savanna trackers, puffing powder or ash to track the slightest air movements. This would have revealed what we know to be true: The air is always on the move.

Walking uphill, we passed through the dark, menacing forms of some old yews, and I suddenly felt the lightest of breezes. This always gives me a pleasant sensation: In woods under thick clouds, a consistent breeze is a friendly arm resting on the shoulder of any natural navigator. It is a prompt to pause and sense deeply, and if we do, another friendly arm reaches out and a long finger points the way.

I asked John if he had picked up the breeze, and he had. We were both sensitive to the wind: It formed a vital part of our personal maps. I paced downhill and uphill repeatedly, gauging the breeze, its direction, character, and strength, until I had found its cause.

The higher we go, the lower the friction that the wind faces and the stronger we should expect it to be. But the way the breeze I felt had suddenly changed made me sure there was more to it than just our altitude. This was a clue to something in our landscape. Finding the exact spot where the breeze materialized, I looked out and saw the map it had made. We had now risen above the main wood below us. Looking to the northeast, the direction the wind was coming from, I could see that although we were still among a mix of yews and beeches, we had risen higher than the tops of the trees in the main wood below us. It was this that gave the wind a chance to reach us. Nothing is random, and this breeze was a strong sign that we had entered a new wind terrain.

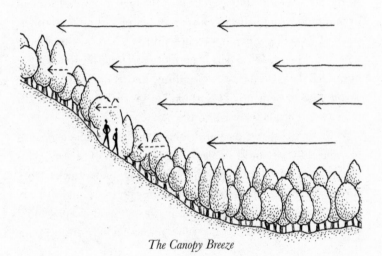

The Canopy Breeze

It was not long before we spotted our first deer, a large herd. Dozens of fallow deer were onto us just before we clocked them and heard the telltale snaps of their hoofs on twigs as they broke away through the undergrowth. A large stag held his ground and stared in our direction for several minutes. We relaxed our statue poses and he melted away beneath the yews.

We walked to the spot where the deer had been grazing, in search of some small clue that might reveal why they had picked that spot on this ridge. They had been grazing among their much-loved brambles, and the nibbled ends were all around. The brambles were thriving in a clearing that let in the light, but there were also dozens of hart's tongue ferns—the only fern in these parts with single fronds: All the others, like bracken, have hundreds of smaller, branching fronds. It was a strange plant to see on top of a chalk hill, surrounded by a grove of beechwood. The hart's tongue favors damp soil and is an indicator of ancient woodland. Its location checked none of those boxes. It was puzzling, but I took some solace from reveling in how it can be used for navigation. The fern is darker in the shade and tends toward a pale yellow-green in bright sun. Darker on north-facing slopes, lighter on southern.

Another crack, and we spotted half a dozen fallow deer on the other side of the clearing. It looked as if we had driven a wedge and parted the herd. The deer would almost certainly circumnavigate us and head toward their brethren. The only question was, which way? They bolted into the wind.

We stood among the ferns and brambles, piecing together the reasons for the deer having chosen this spot. The main building blocks were all in place: There was a good food source in a clearing on the top of a hill that still had tree cover around it, but that much had been clear to John before we set out. Why this hill and not the other? The only logic that came to my mind was that we were downwind of a parking lot, and therefore most human and dog activity, and at a height that the breeze was able to pick up the scent and carry it to the opening. The deer had height and a good scent picture of the nearest threats: humans and dogs. But neither of us could be certain if this was the cause. There are always more windows to peer through.

BLOODHOUNDS

We talked about the way the scent of humans and dogs travels over land toward the deer, which led us to a conversation about the techniques used by search-and-rescue personnel and their extraordinary scent-tracking dogs. John mentioned something that sounded familiar, yet strange. There is a lot of crossover between dog handling and weather signs.

An expert handler of search-and-rescue dogs has to tune in to the most delicate of wind and weather signs—someone's life may depend on a reading of clouds and the faintest breeze. After you have spent time considering the invisible trails that these animals must follow, you're unlikely to think about breezes or look at smoke in the same way again.

The dog handlers' reading of landscape builds on much of what we have investigated so far: radiation, heating, cooling, inversions, laminar flow, turbulence, thermals, and eddies. The sensitivity and skill of those dogs is so far beyond our own abilities that it can be daunting to follow in their tracks, but let us start with the most basic of techniques, one that we weak-nosed humans can emulate in certain conditions.

Let's imagine that there is a casualty—an unconscious person on a hillside. If there is a steady breeze, a dog will need to be downwind of the person to pick up any scent. But where? Neither the dog nor the handler will know exactly, so what can they do to find it? There is no point in going downwind, but there is equally little point in going upwind, which would mean walking parallel to the scent without ever actually picking it up. The only logical thing to do is to track across the wind, to head perpendicular to the wind direction.

Deserts are dry, sterile environments, which means that the scent baseline is very low. Any strong smells stand out, even to humans, and if you walk across the wind for long enough in any desert, you will eventually pick up the scent of smoke. Turn into the wind and track the scent and you will find people sitting around a fire.

So far, so straightforward. But there are times when a dog that is in exactly the right place finds it easy to pick up the scent of a person and times when it is impossible. Humans give off forty thousand microscopic flakes every minute—lovely, I know. Some sink to the ground, but many others are carried in the air. And if you have a tracking dog's sense of smell, those particles stink. The smell is distinctive—the most sensitive dogs can tell the difference between identical twins and pick up illness in a person before they themselves become aware.

If the air carrying this scent passes over a dog's nostrils, it's in business. It turns on a big neon sign in the dog's brain that points the way to the casualty. But if that air misses the dog, then the dog is literally clueless. So, surely all the dog needs to do is to keep tracking across the wind until it picks up the scent.

We have been forgetting one dimension. Sometimes the scent follows the ground all the way to the dog's nose, but at others it climbs and passes over the dog altogether. There are times when a dog can pick up the scent and track it perfectly, only to find it vanishes and reappears in a cycle. The handler has to be able to predict which of these will happen, why, then make decisions about where to lead the dog.

The answers lie in reading the sun, land, clouds, and wind in a way that can only be described as artistry. For us to understand and emulate it, we must switch from the dog's nose to our most powerful sense: sight. Fortunately, smoke behaves in the same way as other airborne particles, like human skin, and acts as a visual marker, allowing us to track the air's gentlest turns with our eyes.

THE SIX SHAPES OF SMOKE

Let's start with the simplest and most intuitive pattern. There is a very light breeze and nothing of note going on in the atmosphere. The smoke will stream downwind, steadily expanding in all directions, in a pattern known as "coning." What does it mean when we see coning smoke? Coning can only happen in a stable atmosphere, and as we know, this means there are no thermals, which in turn means no local heating. There is also no significant cooling from radiation at night. Coning is most likely under skies blanketed with cloud, at day or night. These are great conditions for search-and-rescue dogs.

Now imagine that the clouds clear and the sun breaks through. We know what is going to happen next: There will be localized warming, thermals, and instability. The air rises over some patches and sinks nearby. This sends the smoke on a roller coaster and creates an effect known as "looping."

Sometimes the atmosphere is super-stable: There is a temperature inversion, with warm air sitting on top of cool air. This keeps the smoke from rising or sinking, and it spreads out to form a thin triangle, like a horizontal slice of the cone. This pattern is known as "fanning." It is most likely in the early morning after clear skies overnight. Scent following a fan pattern will pass over any dog that is lower than the source.

A temperature inversion at the start of the day doesn't normally last very long: As the sun warms the land, convection kicks in, the winds pick up, and the layers mix, often leading to looping conditions. But between fanning and looping, something interesting can happen after the sun rises. When the sun is low, it can warm the land weakly, which creates a little instability near the ground but doesn't affect the higher layers. This is a gentle partial stirring of the lowest layer: The smoke cannot rise, but it can and does sink and mix near the ground. It is sandwiched below the "glass ceiling," the top of the inversion layer, and the ground, and sinks to spread evenly in this layer. This effect is known as "fumigating." The spreading makes it easy for dogs to pick up a scent but harder for them to pinpoint its source because the particles have mixed and spread out so much.

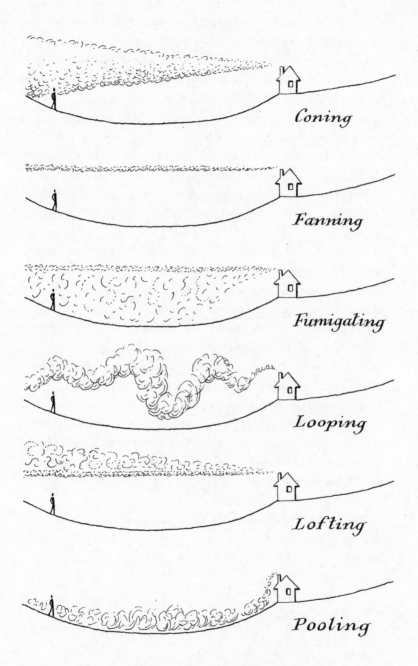

Coning

Fanning

Fumigating

Looping

Lofting

Pooling

After sunset on clear days the heat radiates out of the ground, creating a cool, stable layer near the ground, but without changing the warm, rising air just above it. This leads to "lofting," which can be considered the opposite of fumigating. The smoke will rise and spread widely but not sink below a certain glass-floor level. The only way dogs can pick up a scent in these conditions is to track it down from higher ground.

On a day with few clouds, the sun beats the thermal drum, which leads to a predictable progression of changes in the air from very stable to turbulent. The march is mirrored in the evolving smoke patterns: fanning, then fumigating, looping, and, finally, lofting.

If there is near no breeze, smoke can fall to the lowest point in the local landscape, an effect known as "pooling." This is not useful for scent-tracking: The dog may find the pool, but without a breeze the chain back to the source is broken.

As well as the general atmospheric effects, there are, of course, all the localized ones. Smoke will be seen to eddy if the wind passes any obstacles, and it will rise over hurdles, like trees or buildings, and soar if it touches any strong thermals.

The conditions on our day in the woods lent themselves to coning. Certainly, the deer had no trouble picking up our scent, which could explain their choice of ridge, downwind of the parking lot. I have also enjoyed days when they struggle to pick up our scent even when we're upwind of them, days when we can get almost close enough to touch them.

We put the wind on our right cheek and worked our way down the hill to our parked cars. From here on, let any smells or smoke from any source bring out the bloodhound in you.

CHAPTER 9

Hail and Snow

Blast and Past • Snow from a Blue Sky • Snowflakes •
The Guarantee • The Warm Sign • Accumulation Rules •
The Rebel Dune • Snow Lines

WHENEVER RAIN, snow, sleet, or hail falls from the sky, it is trying to tell us a tale. Hail tells the clearest: All hail falls from a cumulonimbus, the storm cloud. If we see or feel hail, a storm cloud is overhead, the atmosphere is very unstable, and lightning is possible.

Hail forms as water droplets are lifted until they are cold enough to freeze. As small ice particles form, they grow heavy enough to fall with gravity. If nothing stopped them, they would continue falling, then thaw, turning back to water, on their slow journey to the ground. But in cumulonimbus clouds, the ice is stopped on its downward journey by updrafts and sent back up to the higher levels, where more ice bonds to form a bigger ice particle. A small, young hailstone is born. This falls with gravity, a little heavier than before, and, depending on how powerful the updrafts are, it will either fall as a very small hailstone or be raised up once more to grow again and for the cycle to repeat.

The size of the hailstone is therefore an important part of its story. The bigger it is, the stronger the updrafts in the cumulonimbus cloud, the more unstable the air, and the more likely thunderstorms are. The biggest hailstones—and they can grow to a couple of pounds (around 1 kg) in weight or almost 8 inches (20 cm) in diameter—are a sign of a devastatingly powerful storm cloud. In the unlikely event that the size of the hail didn't make you want to run for cover, the storm conditions soon would.

The colors in the hailstone contain their own tale. Hail's journey up and down the cumulonimbus escalator means that it passes from very cold regions at the top down to warmer levels nearer the bottom before rising again. Near the top of the cloud, ice traps air in tiny pockets around the hailstone, giving this layer an opaque white color. Nearer the bottom of the cloud, water forms around the hailstone, which soon freezes. The frozen water has a different look, much closer to transparent. The coats of water and ice alternate and build up, leading to an onion-layer effect in any good-size hailstones. The layers act as a record of the hailstone's journey.

Most individual hailstones are fairly regular, but look at any collection of them on the ground and you'll see a few very odd shapes. Study them more closely and you'll see how two or more hailstones have fused together, creating original and sometimes bizarre forms.

BLAST AND PAST

I recently helped organize a cricket match between "lads and dads": my younger son and his friends versus me and the other fathers. It was late August. The forecasts were so bad that many assumed we'd have to cancel the match. Instead, we changed the plan and decided to play on the beach, not the booked pitch.

We played three hours of cricket in cold, gusting winds and bursts of hail. Each time the hail started bouncing off the sand, the boys and their dads looked around, expecting someone to signal that we needed to stop playing. All I said each time was, "It won't last."

I had some concerns about lightning that kept me vigilant, but the one thing I knew for sure was that the hail would not continue long enough for us to run away from it.

I'm glad we saw it out and finished the match. It was a fun day that the hail didn't ruin. But, more important, the boys were thirteen years old, and we all knew this was the last time we would beat them at cricket.

Since we know that all hail comes from cumuliform clouds, which are isolated, it means hailstorms never last long. There are no blankets of hail. You will have noticed that no sooner have you marveled at the icy balls bouncing off the ground than it's all over. More hail showers may follow, but we have to wait for the next cumulonimbus cloud. If you are sheltering safely and wondering whether to make a dash through hail or wait a few minutes to see if things improve, waiting is a good plan.

The gaps between cumulonimbus clouds lead to extraordinary fluctuations in conditions whenever there is hail. As the towering dark cloud slides over us, light levels drop, winds become gusty, and visibility can deteriorate suddenly. Twenty minutes later the ground around us is covered with icy pellets, but we're standing in sunshine watching the lonely dark behemoth ferry its cold cargo downwind. Hail is theatrical.

Once we have admired the departing cloud's form, it's a good idea to look in the opposite direction. If the conditions are right for one hail-bearing cumulonimbus, they're probably good for several more, and it's only during the gaps that we gain any perspective on what's to come. If the conditions feel safe to do so, the respite between tall clouds is the time to survey the scene and your only chance to gain some warning of another beast on its way.

SNOW FROM A BLUE SKY

You may have had that puzzling experience of watching snow fall from a blue sky. We peer up and there are clouds out there, but none overhead, even as a sprinkling of snow lands on the ground around us.

Bigger raindrops fall faster than small ones because of their greater terminal velocity—their weight-to-friction ratio is better. Snowflakes have a terrible ratio: They are very light and experience high friction. They are the feathers of precipitation, and their journey to the ground is slow enough that their parent cloud can have moved on downwind before the flakes reach us.

The wind can carry snowflakes a long way from that cloud, too—in the high winds of mountainous regions this can be a couple of miles. And some snow clouds will have dissolved before the snow reaches the ground. It never snows *from* a blue sky, but the snow can land on us under a blue sky. The cloud has left the scene.

SNOWFLAKES

Like rain, snow falls steadily from stratus clouds or as showers from cumuliform clouds. In the latter case, it is nearly always from a cumulonimbus. So, if you have snow showers, they can be very heavy, but interspersed with clear skies.

The size of snowflakes is a clue to the temperature. The rough rule is that the larger the snowflake, the warmer the air. The physics is straightforward: Snowflakes that form when it is very cold are made entirely of ice crystals, which are not very sticky. Ice crystals with a thin layer of liquid water coating them are much stickier. If it is warm enough for water to exist in liquid form at the edge of the snowflake, the crystals are covered with a glue that allows lots of layers of ice crystals to stick to each other. The flakes grow bigger.

Snowflakes getting smaller is a sign of falling temperatures. Growing snowflakes signal warming temperatures and can indicate that it may soon stop snowing. This holds true for most blanket snow, but there is an exception to look out for with snow showers. Snowflakes that fall from a single cumulonimbus are sorted by gravity on their way down, and, like rain, the larger ones have a better weight-to-friction ratio and fall faster. In snow showers, it is quite normal to see the flakes getting smaller until they stop altogether.

The exact shape of snowflakes is hard to make out with the naked eye, but there is a beauty to the order. I will mention it briefly here, only because it is reassuring to find patterns and meaning on all scales, even the microscopic.

The shape of snowflakes is governed quite precisely by temperature and the humidity of the air. If the temperature or humidity changes, the way in which the crystals form changes, and the ensuing snowflakes will look quite different. For example, if the temperature stays the same but the humidity increases, the crystals will change from hexagonal columns to hollow columns, then to needles. Or if the humidity stays the same but the temperature drops, needles can change to dendrites.

The most complex, intricate, and beautiful crystals form only in a narrow band of humidity and temperature. But often a simple shape forms in one part of the cloud; this crystal moves to another part of the cloud where the conditions are different and crystals with a new shape are added. Thin columns are given wide "hats" at either end. We could, in theory, with the aid of a good microscope in a cold room, study all of the different crystals in the snowflake and write the story of its journey around that cloud.

THE GUARANTEE

Snow is, of course, more likely in winter than summer, but in many places more likely in early spring than the dead of winter, for two reasons. First, the sea and the land are colder in late winter and early spring than they are in midwinter. Second, snow owes as much to the air masses as to the season, and air that has arrived after a journey over frozen continents is more likely in many places in early spring. A cold air mass can lead to snow in June, but you cannot have snow at any time of year in a warm air mass.

Predicting if and exactly when snow will fall is challenging. We have all the same requirements for rain, but the need for quite specific temperatures, too. Snow is really fussy about temperature—a couple of degrees changes the likelihood, quantity, and form of snow. This means that a few hundred feet of altitude can change things entirely. In honesty, sometimes the best we can do is spot the difference between the blankets and the showers. As you'd expect, blanket snow is likely to continue for hours; showers will be intermittent and potentially heavy. Both will be strongly influenced by the way they are carried over mountains, in very similar ways to rain.

There is another way of looking at snow's sensitivity to temperature. In many landscapes it will be hard to predict whether snow will fall in an exact spot because it may be a degree too warm or cold. But in the mountains, we can turn that on its head. In mountain ranges there is a sliding scale of temperatures with altitude. If it is a little too warm for snow where you are, it will rain, but it will snow a few hundred feet higher, and vice versa. In low country, we're trying to determine whether all the pieces of the jigsaw are right for snow. But in the mountains, once we have figured out that precipitation is on its way, we know there will be snow: It's just a question of altitude.

THE WARM SIGN

The coldest weather and snowfall rarely come together. As we have seen, the lowest temperatures occur when the skies are clear, typically under high-pressure systems at night. There will be no snow under completely clear skies.

In winter it is more likely that snow will be signaled by temperatures rising after a cold snap. This is what lies behind the old lore, "When the icy wind warms, expect snow storms." It is a little counterintuitive, but stepping stones lead to the logic: Temperatures rise when clouds cover the land; clouds replacing clear skies signal a front; fronts bring precipitation; during a cold winter spell this increases the chance of snow. As with all fronts, monitoring the progression of cloud types and wind shifts will complete the picture.

ACCUMULATION RULES

Once snow has accumulated, it begins a new life, and we find a world of patterns to look for. Why has it accumulated over there and not here? For snow to stick, the ground must be cold enough, and, as we have seen, the land can be warmed in several ways, but mainly from below and above. Snow is sensitive to temperature in many of the same ways that frost is, but with a few of its own habits.

Snow is more likely to accumulate on grass than on soil, as soil is a better conductor and draws more of the earth's heat from below. Pavement conducts even more heat, so it's even less friendly to snow. You may notice that bridges and overpasses allow snow to accumulate before the lower roads nearby—bridges are suspended and cannot conduct heat up from the ground. This is why it is common to see signs that say, "Bridge Freezes Before Road."

Overpasses are normally snowier than bridges over rivers, as the air above rivers will be warmed by the water and rise to the bridge.

The dark surface of roads absorbs more of the sun's warming radiation (even on cloudy days) than either grass or soil. And cars add more warmth still. That's why snow sticks to lighter-colored pavements before it has accumulated on roads, and pavements on bridges are snowier still. In summary, there is an order to the places where snow sticks, and by the time roads have accumulated snow on them, the land on either side will have a thick layer.

The size of snowflakes will shape how they behave once they land on the ground. Large snowflakes tend to be sticky (with liquid water around their edges) so they bind with other snowflakes on the ground, and will stick if the ground is cold enough. Small snowflakes are drier, don't stick to each other, and can be blown by the wind after they have landed. You may have seen this dry snow—it looks a little like Styrofoam flecks moving on with every puff of wind.

THE REBEL DUNE

If it's windy and there's enough dry snow, new patterns start to form in the landscape. The wind carries dry snow and it keeps moving until it is stopped by some barrier. Then it accumulates as a snowdrift.

There is an art to reading snowdrifts, and it's easier to appreciate once you know some of the shapes to look for. Readers of my earlier books will be familiar with one in particular. The wind sculpts a similar shape in waves, sand dunes, and snow. In each case it creates dunes with a shallow side and a steep side. The shallow side forms on the side the wind has blown from and the steep side forms on the side the wind is blowing toward. In sandy or snowy environments, these dunes vary in height from inches to hundreds of yards and can be found in the thousands.

Once you have spotted a few, it's worth keeping your eyes peeled for a less well-known shape. I call it the "rebel dune," and you'll find it in places where there is a short vertical wind barrier, like a fence or hedge.

The wind blows across the landscape, creating dunes that follow the rules, shallower on the windward and steeper on the downwind side. Then the wind hits a fence or other barrier. Suddenly it's tripped up. Eddies form on the downwind side of this barrier and the wind curls down and starts blowing the "wrong" way, back toward the barrier—against the main wind. This rebel wind picks up some of the loose snow and starts building its own dune, a rebel dune: It points the "wrong way." When you know to look for it, you will spot a few large examples, but with practice it will start to appear before your eyes on a much smaller scale. I have seen tiny rebel dunes on the downwind side of small stones, furrows, and even frozen horse manure.

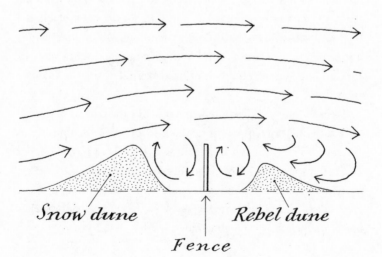

Snow dune   *Rebel dune*

Fence

On mountains, snow is always on the move. Mountains and valleys form a mix of exposed and sheltered environments on steep slopes. The wind transports snow from exposed areas and deposits it in sheltered ones, with a tendency to carry it downhill. Even when snow thaws and refreezes as ice, then clings to a valley as a glacier, it is still on the move.

In winter, in exposed mountainous areas, snowdrifts collect on the downwind side of rocks, ledges, and bushes. In spring they thaw, which leads to moister ground. Many mountainous regions are arid, and those moist pockets are havens for spring flowers and other plants that might struggle without the watering effect of the snow.

SNOW LINES

The snow line, the line where accumulated snow stops and gives way to clearer ground below it, undulates around each mountain. It is shaped by aspect, seasons, and the climate, being higher on south-facing slopes and many thousands of feet higher in the tropics than it is in temperate zones. You can step onto snow from the sea in the Antarctic but may have to climb for days to find it in the tropics. It also varies with the microclimate. Warm local winds are funneled over and around high ground, devouring snow as they go and leaving channels of bare ground jutting up into the snow line.

Trees in snowy and exposed climates, like the pines of Lapland, have different branch patterns, which reveal their relationship to the snow. The branches buried under the winter snow actually fare better than those above it because the higher ones are ravaged by winter winds and the ice crystals they carry. This creates a layered effect, and in summer the robust, healthy lower branches reveal the height of snow in winter.

In many regions, organisms paint tree bark and rocks with clues to the height of winter snow. For centuries, mountain residents have observed how some lichens grow above the snow line on the trunk and stop abruptly where they meet the top of the winter snow. The best-known snow-gauging lichen is *Parmelia olivacea*, which grows on the bark of mountain birch. It has been used and noted by scientists for almost a century, and by mountain folk for a lot longer.

Many mountain plants depend on the blanket that snow forms: It keeps heat from escaping as radiation and traps it nearer the surface. Or, as the Russians would put it, "Corn is as comfortable under the snow as an old man is under his fur cloak." In the mighty, wild primeval Białowieża Forest in Poland, ivy on the trunk marks the average high point that the snow reaches. Animals, such as grouse and hares—and some intrepid humans, too—use the warm blanket that snow offers for survival.

The snow blanket leads to one surprising and potentially hazardous situation. If you find ice over a pond or lake that is thick enough to walk across, beware of any snowy patches. Ice can thaw quickly under snow's cover and become dangerous before the bare ice that surrounds it.

Every plant is a specialist, and in the mountains each one reflects a different relationship to snow. On the lowest slopes, some plants will die if there is more than a little snow; much higher up, some mosses cope with only three snow-free weeks of the year. And some ridiculously hardy plants, like the glacier buttercup, have evolved to fight their way up through snow.

In these exposed areas, we can read every plant as a clue to how long the snow season is in that spot, because each species needs a certain minimum amount of snow-free time. Dwarf willow, for example, is a sign that we can expect at least two months without snow cover each year.

CHAPTER 10

Fog

Sensitivity, Near and Far • The Big Two • Look for Dew •
Water Maps • Fog Is a Forecast • Advection Fog • Upslope Fog •
Steam Fog • Haze • Fronts and Visibility

ON DECEMBER 11, 1990, a highway patrolman drove along I-75 near Calhoun, Tennessee. It was a stretch of road known for problems with fog, but he didn't see any so did not activate the flashing lights to warn drivers of poor visibility.

A little over an hour later, there was a car crash. A vehicle had slowed down in thick fog but the one behind it didn't. The second piled into the first. Amazingly, both drivers were uninjured and climbed out of their vehicles to assess the damage. Seconds later a car smashed into the back of the second stationary vehicle. A fire burst out of the wreckage and consumed all three vehicles.

On the opposite side of the road a car entered the thick fog and slowed; the one behind it did not and smashed into the back of the first. Then three more vehicles, including a pickup truck, crashed into them.

In total, ninety-nine vehicles collided with each other that morning in Tennessee. There were twelve deaths, and fifty people were injured. There was one positive outgrowth of this horrific accident. The number of casualties meant that there had to be a thorough investigation. Authorities needed to understand the cause to save lives in the future. Why did so many die that day?

Investigators knew that fog was at the center of the puzzle. But there was often fog on that stretch, yet nothing on this scale had happened before. There were suspicions that a nearby paper mill might be partially responsible. The plant included wastewater ponds, and one straddled the highway near the accident.

A thorough inquiry grappled with all the observations and available data. The experts used computer models and the latest science to assess the cause of the particular marshmallow-like fog. It went to court, where the paper mill settled and closed one of the ponds, even though they were found by one investigator not to have made "a significant contribution to the fog."

The intricacies of fog, like all weather, are complex. But in the report of that accident, there is some clarity. Fog was common on that stretch of the interstate, and there had been six previous serious accidents, including three fatalities, in 1974 and 1979. Each of the seven accidents had something in common. They had all happened in fog that developed shortly after sunrise on calm, cool mornings, following warm days and a clear night in an area with lots of water. Another familiar recipe: air, water, and heat, mixed in a particular way.

SENSITIVITY, NEAR AND FAR

We all know when we're in a thick fog, but there are many shades between perfect visibility and whiteout. We can sharpen our sensitivity to them by learning to look for the same known features on a regular basis, some nearby, others far away.

If you are fortunate enough to have access to good views, pick a series of landmarks at varying ranges and make a mental note of when they are clear, faded, and invisible. In cities you can get good views from higher buildings but also from the center of bridges. If it is hard to find one, try to settle on an object in the middle distance with lots of detail. Can you recognize faces or see the colors of people's clothes when they're at the end of a long street? The answers change according to the visibility.

Again, feel free to cheat: The next time you know a period of fine, sunny weather is about to break, keep a vigil on distant landmarks. Watch the details fade.

THE BIG TWO

When the air can hold no more water as gas, it is saturated. If we add more water or if the temperature drops, the water vapor condenses and tiny droplets hang suspended in the air. They scatter light, which reduces visibility, in all directions, giving the air a white appearance. Reduced visibility with a white tinge has two familiar names: mist and fog.

Mist and fog are the same phenomenon; fog is just the more extreme version. In other words, mist is half-hearted fog. From here I'll refer to both as fog, because I don't want to patronize it.

Like precipitation, there are lots of names for fog, and plenty of nuances and local subtleties. But we can understand most of what we need to know about it by recognizing the two main characters: radiation fog and advection fog. In the simplest terms, radiation fog forms over land and advection over sea. Here is another helpful diagnostic tool: If you see a fog first thing in the morning, the finger of suspicion must point at radiation fog. If it forms during the day, the blame shifts to advection fog. And once we have the main suspect in our sights, we can see how it fits into the scene and draw other useful conclusions.

LOOK FOR DEW

Radiation fog is the one you are most likely to see, partly because it's the most common and also because it favors land, where we spend most of our time. It forms in similar conditions and in a similar way to dew, and that is one of the first things we can look for.

There will always be dew with radiation fog. Like dew, radiation fog forms when heat radiates out of the land. The layer of air touching the land is cooled by it and its temperature falls below the dew point. This typically happens overnight, under clear skies, and with low or no winds—classic high-pressure conditions. It is most common in autumn. So far, the formation of radiation fog is following that of dew, but fog needs more water in the air than dew. It needs very wet air indeed. So our first question, on spotting fog, is to ask, "Why is the air so wet?"

There are only two places the moisture can have come from. It was either brought in from elsewhere by the wind or it has come up from the ground. But we've established that this is a radiation fog, so the air is calm. Winds can't be the cause. This means the moisture has to have come from the ground.

WATER MAPS

Radiation fogs are much more common when the ground is unusually wet, especially after heavy rains, when it is sodden. Clear skies overnight after hours of frontal rain set the stage nicely. But it's more interesting when we can't point the finger at rain-soaked land: It suggests the water has come from somewhere else. Radiation fogs can map rivers, lakes, and floodplains for us. If the weather has been dry for a few days and you are traveling through a landscape on an early morning and suddenly encounter fog, the chances are that you are very near a large body of water.

A train I catch regularly takes me from the Sussex town of Arundel to London. Like most railway lines, it has to perform a balancing act. It can't go over high ground and instead must hug the valleys, but it can't leave itself vulnerable to the river floods either. The metal ribbon snakes along an embankment, sometimes almost touching the river. On radiation-fog mornings, the river is mirrored by a fatter serpent of fog, which betrays the location of the water below. I like to watch as a white curtain lifts and falls as the train weaves in and out of the curved fog bank.

FOG IS A FORECAST

Radiation fog looks and feels wet, and it is, so it can instinctively feel like a bad-weather sign. But it forms under clear skies so is a sign of good weather to come. Our old friend the Greek philosopher Theophrastus wrote, "Whenever there is fog, there is little or no rain."

When we walk out into a morning fog, it can be hard to imagine that the sun is still burning warm and bright, and the sky may be clear only a few dozen feet above our heads. And all the while it is burning off the water. Fogs shrink from the outside in, like a piece of paper burning from the edges. Fog banks also grow shallower as the sun rises, so make sure you look up.

One of the first signs of a fog about to clear is that blue sky can be seen directly over your head. The next time you look down into a valley and see a thick fog there, take a moment to imagine you are inside it. Feel the warmth of the sun on your face and neck. Now share a shiver for the valley folk below you as they stumble around barely able to see each other.

The time of year will have a big impact on how a radiation fog behaves. Since the sun banishes the fog, it follows that it lifts quickly in summer but lingers longer in winter, when the sun is much lower in the sky and weaker. Hence the two old lore sayings, "A summer fog is for fair weather," and the chillier, and more fun, "A winter fog will freeze a dog."

Once the sun has thinned the fog, more of its warming rays reach the land, and that, too, begins to heat. The air just above the ground begins to warm, expand, and rise, lifting the mist with it. A rising mist is a good sign: It won't be long before you feel the full heat of the sun.

You may hear people refer to a "valley fog." This is just a particular type of radiation fog. Think back to the frost hollows: Cold air sinks into valleys, making the conditions ripe for dew, frost, and fog.

Each time you see a morning fog, I'd encourage you to think of it as a double sign: good weather in the near past and in the near future. Recall clear skies of the night before and look forward to good weather later in the day. Now see if you can find the water causing the fog. Have there been recent heavy rains? If not, try to identify a nearby body of water.

ADVECTION FOG

If you see a fog develop during the day and can feel the wind, your chief suspect is advection fog (advection means horizontal movement of heat). Like radiation fog, advection forms when the temperature of the air drops to the dew point, but unlike the still-air radiation fog, advection fog forms when very moist air is blown over a cold surface.

Imagine the spot where two sea currents meet. One is very warm and the other is very cold. If a wind blows from the first toward the second, then warm, very moist air will be carried over the cold surface of the second. The vapor will condense and a large, dense fog bank will form. Regular thick fogs occur where the cold Oyashio current meets the much warmer Kuroshio current off the eastern coast of Japan, and off Newfoundland where the warm Gulf Stream and the cold Labrador current collide.

Exactly the same process will happen if a warm maritime air mass is blown over cold water. I found myself in this situation near the Channel Islands off the coast of France. An abnormally warm, moist August air mass was cooled by the sea, and we were sailing in strong winds and thick fog. In the testing conditions my friend lost his signet ring overboard into the deep, never to be seen again. At the time I was too busy to give it much thought and consoled my friend. Recalling it years later, I wondered if the weather gods had demanded their tribute and received it. We were sailing near treacherous rocks and battling fast currents, strong winds, and terrible visibility—in short, potentially hazardous conditions. If the ring had not been surrendered, what price might they have exacted?

If there are onshore winds, the sea fogs are blown over land and are well-known characters in many parts of the world, like the haar in northeast England or along the Pacific coast of the US. The fogs don't travel far inland, but they bring much-needed moisture to some dry coastal regions and can sustain species in these arid belts. In extreme cases, the meeting of very wet air with very dry land creates a unique ecosystem known as a "fog desert." The Namib, Atacama, and Baja California deserts all include a belt.

Fogs can add a lot of moisture to the ground even in temperate zones, and you may have heard this process in action. "Fog drip" is the name given to the fog-fed droplets that collect on tree leaves and fall to the forest floor.

Trees change fog. Forests act as fog-catchers and have been planted in places like Japan, where sea fogs cause problems in coastal areas. The fog harvests some of the water, as above, but it also traps some of the tiny airborne particles that encourage the fog's condensation. If the breeze is right, you may find fog-free pockets just downwind of woodland, but if there is no wind, fogs can linger in the cool woodland for longer than they do outside.

UPSLOPE FOG

If a wind carries moist air far enough uphill, it will condense and form fog. It forms on all mountains, on their windward side, when the conditions are right, surprising walkers regularly on humid sunny days, when it seems there isn't a hazard in the world. Then, like some fairytale monster, a fog from nowhere climbs up the hillside and gobbles you up.

The process is nearly identical to how we saw clouds form on the windward side and summit of hills and is just a question of perspective. If you're high up a mountainside and a wind envelops you in a white blanket, you will call it fog. If you are below that blanket, you might call it a cloud. And if you're above it, it will answer to either name.

If you are heading up a mountain, stay tuned to all the signs we have looked at, including the seven golden cloud patterns. If the clouds are lowering, humidity is rising and the risk of fog forming increases. You may be engulfed by a fog from below, long before the higher clouds drop to your level.

STEAM FOG

When very cold dry air is lying over warm seas, water vapor evaporates from the sea, mixes with the colder air just above, and instantly condenses to steam. It snakes upward, more like smoke, hence its nickname, Arctic sea smoke. If the cold air was moist, not dry, advection fog would form, but because the air is dry, the fog soon dissolves and disappears.

It is always eerie, finding yourself bobbing on a boat as the sea appears to smolder all around. But you don't need to go to sea to witness this: You will have seen steam snakes rising from ponds, lakes, and rivers, especially in autumn when the air loses its heat before the water does. It is the same phenomenon as the steam above a mug of tea on a cold day.

HAZE

The meaning of "haze" is hazy. It means different things to different people. The most common meaning is: visibility obscured by dry particles suspended in the atmosphere. They may be from smoke, harvest dust, trees, or many other sources. Haze dulls and changes the colors in the landscape, often giving the air a brown, blue, or yellowish tint. The exact effect depends on the nature of the particles, the background, and where you are looking relative to the sun. Haze is normally a summer phenomenon, seen during warm days and light winds, and can be a sign of a temperature inversion.

FRONTS AND VISIBILITY

As a warm front approaches, the cloud base drops steadily and rain falls into the air below. The air may become saturated, at which point we will experience rain, fog, and wind.

Even when no fog is running before a front, every front will still bring a roller coaster of changes in visibility. The standard progression looks like this:

Stage	Visibility
As the warm front approaches	Good to start with, then steadily deteriorating
At the warm front	Poor
In the warm sector	Very poor to foggy
At the cold front	Poor
After the cold front has passed	Excellent

The crystal-clear air after the cold front has gone through lifts moods and offers us an opportunity to admire the mighty towers and thunderstorm clouds as they threaten the land downwind of us.

Rain at any time reduces visibility as it falls but often leads to an improvement afterward. It clears the air of dust, pollutants, and other particles. This effect is most noticeable in the warm, sultry days of late summer when heavy showers break a heatwave and scour the air. But if visibility remains poor after rain, there's more to come.

Many fogs form thanks to a mix of the processes above. A radiation fog that forms in a valley may be lifted by a breeze and pop up over a ridge. In the Sussex countryside these roaming, climbing fogs were once known as "call boys," but I nickname them "X-ray fogs." In the hills where I live, they pop up like signs, pointing to the valley on the other side of the high ground.

The next time you see a fog, try to identify the main cause as radiation, advection, or a front. Then give careful consideration to the season, any wind, the shape of the land, and your altitude, as this will inform your experience and tell you what to expect.

Your feelings about any fog will be tied to how fast you're moving. If you're standing still, they're friendly puzzles, waiting to be solved using the techniques above. If you're walking up a mountain or traveling in a car, they call for caution. Every fog is one more good excuse to pause, tune our senses, and solve a small mystery.

CHAPTER 11

Cloud Secrets

IN JUNE 2019 I was fortunate enough to be shown around the new Defence and National Rehabilitation Centre at East Leake, Nottinghamshire. The facility is extraordinary, complete with virtual-reality simulators and the latest high-tech equipment for helping injured military personnel with their recovery.

The enormous redbrick building was daunting from the outside. I was given a tour by Luke Wigman, a Royal Air Force serviceman who had lost a leg after stepping on an IED in Afghanistan five years earlier. To be honest, I found the combination of trauma tales, recovery, and technology a little overwhelming. I had no doubt about the brilliance of the facility or the courage and dedication of those in it, but I struggle with medical institutions. I feel queasy going into a hospital as a visitor.

We stepped outside, and I walked off the path onto the mown grass. I asked whether green open spaces were important for mental health and recovery. I was interested to know, but also happy to be wandering around the grounds, back in terrain I could relate to. I listened as Luke explained the benefits of a rural setting. Then I saw a cloud in the distance. It relaxed me and made me smile. I asked Luke if there was a town in that direction.

"Loughborough's over there." He looked a little puzzled. I explained that an isolated substantial cumulus cloud that hovered over the same spot and didn't move as other, smaller, clouds drifted with the wind meant there was a powerful source of heat beneath it. It was early afternoon in summer; we were surrounded by green fields in a heavily populated part of the world; the cause was easy to determine.

Luke was polite and showed interest, but I didn't expect him to care. And I didn't care if he cared: I just wanted another minute outside, and talking about clouds was the only way I could think to engineer it.

SEEING CIRRUS

We are now comfortable with cloud shapes, and there are only three main ways to describe them. They are heaped (the cumulus family), layered (the stratus family), or wispy (the cirrus family). All clouds fit one of these categories or a combination of them.

Clouds in the stratus family carry signs, but their blanket nature means they are nearly always telling us that more of the same is to come, or that the change will be glacial in pace.

The real joy of finding signs in clouds comes from getting to know the cumulus and cirrus families well. Cumulus clouds have featured strongly so far, and we will meet them again many times, but we will spend most of this chapter with the cirrus family.

Imagine that three people have returned home from a walk. They all followed the same path, but they did not have the same experience. The first person enjoyed the time when the path snaked alongside the river. The second person saw where the rough, riffling water suddenly calmed at a deep pool in the sunshine and made a mental note to come back for a swim on a hot day.

The third, sensing a cloud pass in front of the sun, knew this would mean insects falling from the air onto the water. They looked at the way the water ran fast over the rocks but slowed near the pool and snaked around the jutting rock at just the right speed. This was where the bubbles were congregating and where the fallen insects would, too. This was also where the trout would rise to the surface to feed—and, yes, there it was! The tiny circular ripples of the "kiss" where the fish's mouth took the insect. There is seeing and seeing.

We all see the sky. Some of us notice the wispy clouds that scratch its surface. We will be the third person. The cirrus clouds are the trout's kiss on the blue sky, intricate traces that reveal a richer story.

CIRRUS COMMAS

Cirrus clouds, like ripples on the surface of water, come in so many shapes and patterns that it's tempting to see them as random, little more than visual noise. But they can't be random: There is logic behind every single strand and shape.

Like most clouds, cirrus are more common over land than sea. Logically this can only mean that the very thin, hairlike strands that float more than twenty thousand feet in the sky are shaped by our landscapes in some way. However, unlike the much lower cumulus clouds, it is hard to link individual cirrus clouds to exact landscape features. They are more common over mountain ranges, but instead of trying to map the land using cirrus clouds, a tough task, our focus shifts. Our time is much better spent using these high clouds to glean clues about moisture levels and what the winds are up to. And to do this we need to appreciate that every wisp, every strand is a written line. Learning to read the cirrus script starts with a simple shape: the "comma."

Cirrus clouds are always made of ice crystals; they are far too high and cold for water to exist in liquid form. They begin life at a "head," the spot where the ice crystals are forming in their millions, which acts as the source of the cloud. As soon as the crystals have formed, they begin to fall and leave a trail. The head and the trail together look a little like a thin white "comma" in the sky. It can be a squashed, twisted, or backward comma—this is not a cloud that cares much for formal punctuation.

The official name of the cloud in Latin is *Cirrus uncinus*, which means "hooked" cirrus. The key to recognizing it is a thicker, denser tuft of cloud with thinner hairs streaming down from it. Study cirrus clouds over the course of a few weeks and you are guaranteed to see and recognize your share of commas. They are common and unmistakable when you know what to look for.

Once we have spotted our comma in the sky, we're ready to find the sign in it. The ice crystals fall into the air below the head and evaporate long before reaching the ground, but before they disappear, they leave lines that we can see. These white trails are known as fallstreaks.

If the air below the head were moving at the same speed and in exactly the same way as the ice crystals in the cirrus head, we would see vertical, perfectly straight fallstreak lines. But this is very rarely the case. Wind speeds and directions nearly always change with altitude. So, we find that the fallstreaks are mapping the difference in wind speed and direction between the highest part of the cloud—the head—and the air below it.

Look for this the next time you see a cirrus comma and you will spot the two main parts, the head and the fallstreaks, and you will be able to decipher what is going on.

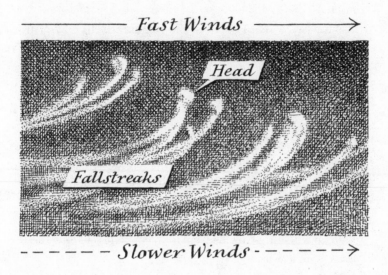

The comma has shown us which way the upper wind is blowing. This can be useful for a number of reasons. The upper winds can be used to navigate. They don't offer precision, but there are some situations—deep in a wooded valley, for example—where you can see high clouds, but not the sun or other indicators. Although these winds make an imprecise compass, they hold their trends for a lot longer than the lower winds and clouds. If the upper clouds were moving from west to east in the morning, it's very probable that they're on the same course at the end of the day.

MARES' TAILS

The cirrus comma also offers a very simple forecasting method. As we have seen, the closer to vertical the fallstreaks, the gentler the change in wind speed and direction with altitude. If the fallstreaks are in the same direction as the head is moving and have a very slight bend, it's a sign that good weather will continue.

Conversely, a sharp curve in the comma reveals an equally sudden change in wind speed and direction just below the head. A sharp change like that is an example of wind shear and is a strong sign that bad weather is on the way. These sharply bent cirrus clouds have been nicknamed "mares' tails," and this is the logic behind the traditional belief that mares' tails are a portent of worsening weather.

We can also use the comma to compare the wind direction at the level of the cirrus with the wind direction lower in the atmosphere. If the upper winds are blowing in the same direction as the lower main wind, then good weather should continue. If the high and main winds are blowing from markedly different directions, change is on its way.

One comma on its own may not signify anything other than local wind conditions, but if we see several contorted commas and more forming by the hour, then we have both wind shear and rising levels of moisture in the upper atmosphere: A low-pressure system is approaching and bad weather is about sixteen hours away.

JET-STREAM ROPES

In August 1947, back when passenger aircraft still had names, an airliner called *Star Dust* took off from Buenos Aires, Argentina, bound for Santiago, Chile. The route requires a climb over the Andes Mountains before descending on the Chilean side of the ridge.

In the age before GPS, a method called dead reckoning was vital to navigation, for walkers, sailors, and pilots. The simple principle is that if you know the speed, direction, and time you have been traveling, you can work out where you are relative to your start point. In the simplest terms, if you start at your home and walk north at three miles per hour, one hour later you will be three miles north of home.

For more than a thousand years this simple method lay behind all navigation in terrain without landmarks. If you sail west from Ireland at five knots for three weeks, you will be getting close to the Americas. Start point, direction, speed, and time. What could go wrong? The one seemingly small thing that might throw you off is sea currents you're not aware of. A current of only one knot flowing south would send you more than 10 degrees off course, about five hundred miles south of where you thought you would be in the case of an Atlantic crossing.

The same principle is true in aviation, but here we call the current "wind." The pilots of *Star Dust* were very experienced— all three had flown combat missions during the Second World War—and they did what all professional pilots did in their situation before any flight. They planned the direction, speed, and time they would fly, took the forecasted wind into account, and set off.

A few hours later the aircraft disappeared and nothing was seen of it for decades. Conspiracy theorists had all sorts of fun: Clearly, evil was at work behind this mysterious vanishing.

In 1998, more than half a century after the aircraft had taken off, two Argentinian climbers were on a glacier fifteen thousand feet up in the Andes, when they noticed pieces of an aircraft engine protruding from the ice. They had found the remains of

Star Dust. Conspiracy theorists don't give up lightly, and theories about the cause of the accident continue to swirl about the Andean peaks. But there is now a consensus about what happened that day back in 1947. The flight was doomed by a wind that the crew knew little about: the jet stream.

The science of jet streams was in its infancy when *Star Dust* took off, and the pilots had not included it in their thinking. They had flown into a headwind far stronger than they could have guessed. Cloud cover meant that they couldn't see exactly where they were, but that was quite normal, and they used dead reckoning to estimate their position, as they had done hundreds of times before. When their calculations told them that they were safely past the Andean summits, they began their descent toward Santiago. Unfortunately, the powerful headwinds meant they were well short of where they believed they were. They hadn't flown over the summits yet, and *Star Dust* descended into the mountains on the Argentinian side of the ridge.

Jet streams are snaking, high-speed, high-altitude winds. On average, they flow from west to east and are like rapid aerial rivers: broad, shallow, and faster in the middle than at the edges. They form at the boundary of very cold and warm air masses. There are two jet streams in the northern hemisphere: the subtropical jet and the polar jet.

The exact anatomy and behavior of jet streams is complex, and we don't need to study it to read the signs these winds create. In truth, the science is still young, and meteorologists continue to debate the intricacies. Does the moon have a significant impact? Maybe. We'll leave them to their machinations, as the fundamentals are now well understood and that is all we need.

The position of the jet stream fluctuates, moving north or south, as it flows from west to east. And it snakes on its

Cumulus mapping the Isle of Wight and cirrus building ahead of a front

Hoar frost forms on the low grass and leaves—but not on the track, the vertical stems, or under the cover of the wooden rail.

The thistle stops heat rising from the ground better than grass and is much colder.

Rime ice forms after a cold fog blew from the north. We are looking west.

A rock conducts heat from below and melts the snow.

A shower cloud has formed over a dark wood on a hill. The pannus reveals that the shower has started.

A cliff cloud over the Isle of Wight

Fanning smoke and mist over a village during a temperature inversion: The weather felt very different above and below this line.

The southern sun has melted the snow on one side only. We are looking west.

Radiation fog is forming, but only under clear skies and not under clouds or trees.

Fog drip: The fog is mapping a hidden river.

An anomalous cumulus cloud maps the hidden town of Loughborough.

Cirrus commas—"mares' tails"—over Brighton: A front arrived the following morning.

Jet-stream ropes and weak halo: Rain started twelve hours later.

The "tall tell": Cirrus above stratus reveals hidden storms.

Fat, spreading contrails over London: The good weather ended the following day.

journey; it may flow from south-southeast to north-northwest, or south-southwest to north-northeast, but it tends toward a west-to-east path.

The position of the jet stream will always reveal something about the weather we will experience in the northern temperate zone. We can think of the jet stream as a line that runs between two different weather zones. It follows that if this line moves over us, we must be sitting under the boundary of the two zones. The weather is unlikely to be settled. Spotting signs of the jet-stream winds is therefore an early clue that low-pressure systems and the bad weather they bring may be on the way. In fact, it is one of the earliest such signs.

The jet stream blows miles above us, and, like all winds, it's invisible. But it does leave traces that we can spot in high clouds. The jet stream creates its own brand of cirrus clouds. Individual patches of cirrus can reveal the wind direction, and, as we saw in the comma cirrus, they can also map local changes in wind speed and direction. But when the jet stream is overhead, the cirrus picture changes dramatically and we start to see very long, linear cirrus patterns in the sky. I call them the "jet-stream ropes." They are so easy to spot when you know to look for them, but equally easy to miss; cirrus is always so high and thin that even when it covers large parts of the sky, it never blots out the sun or casts a shadow.

If you see cirrus that forms very long lines that can be followed across the sky, often all the way from one horizon to the other, you are looking at jet-stream cirrus. All sorts of patterns can be superimposed on the main linear nature of these clouds—fishbones jutting out of the spine is a favorite. Sometimes there is one strong, broad backbone of cloud, sometimes several thick bands, and at other times it will be dozens of very thin lines. But, whatever secondary patterns you spot, the key indicator is that the ropes run across most of the sky.

Since the clouds form at heads, they can also look like super-stretched commas, but covering much larger areas of the sky. A cirrus comma won't be much bigger than an outstretched fist (approximately 10 degrees), whereas ropes typically stretch across more than half the sky. The ropes don't have to be perfectly parallel, but they are all being pulled by the same fast, high wind, so there will be a definite trend in one direction. This will average a west-to-east orientation but, like the wind itself, they can move quite far from that.

The other thing to look for is their speed. All cirrus clouds *appear* to move quite slowly, but this is deceptive and is due to their height. A car doing 100 mph will appear slow if it's on a road on a distant mountainside. Cirrus clouds all move at the speed of the strong winds at that high level of the atmosphere, but those winds fluctuate in speed and are noticeably faster when the jet stream is in town.

It is difficult to gauge the speed of cirrus clouds by looking at the clouds alone; we need something stationary to compare them to. They are nearly transparent, so we can use something in the foreground or behind them; either will work, as long as it appears stationary. The moon or stars are good, but the sun is too bright to use. Find a jutting high branch of a tree (a thick one, not waving leaves or twigs), power lines, or even the corners of high buildings in cities. It's really important to focus on an exact part of the cloud. Try not to gauge the general movement of a large body of cloud, but pick one feature and track it. This is the only way to be accurate.

When viewed in this way, most cirrus clouds, like the comma, appear to move slowly in a constant direction. But once you're used to this default slow speed, you will soon notice that the jet-stream ropes don't linger or dawdle. They are among the fastest-moving clouds you will ever see, and even though they are the car on the distant mountain, we can still sense the jet stream's engine roar in the way they move past our reference point.

What do the ropes mean? They indicate that the jet stream is almost overhead. This means that an increase in winds is likely in the next twelve hours and a low-pressure system and the associated fronts will be close behind. Expect a warm front within twenty-four hours. The sign is especially strong if the alignment of the ropes is from northwest to southeast.

CROSSHATCHING

After spending time watching cirrus clouds move and gauging their speed, it won't be long before you find a skyscape with two distinct layers of cirrus clouds. This is a gift, a golden opportunity to map the winds at these two levels.

We can use the methods above on each cloud within each layer, and this will give you a good picture of their movement. But you can get an instant snapshot in the shapes you see, too. Are the lines parallel or crossed? Our eyes will quickly pick out crossed lines, which is an instant clue to a significant change in wind direction between the two layers. If there is a marked difference, the effect looks like crosshatching.

The sign here is just a clearer, easier version of the cirrus comma. If the winds are moving in the same direction, it means good weather will continue. If there is a marked difference, deterioration is likely. "If you see the crosshatch, soon time to close the hatch."

COLD-FRONT CIRRUS

The tops of storm clouds are ice. As the mighty cumulonimbus clouds mature and then decay, the icy tops may break off and form independent cirrus cloud formations. They are denser, much more like cotton candy, than most cirrus and can be more persistent than the storm clouds. They like to linger long after their parent cloud has dissolved, and will be seen a long way downwind from that storm.

Cold fronts are notorious for their turbulent, unstable nature. Even when storms don't materialize, the conditions are still right for very tall cloud formation. The taller the clouds, the more likely they will reach icing levels. It's common to see cirrus formations after the cold front has passed.

It's worth remembering that cold fronts and storms leave cirrus clouds in their wake, otherwise it's tempting to see them as a portent of what is to come. Once you recognize them as the remnants of larger, older clouds, it's easy to see that they're only a sign of what has passed and don't warn of more storms to come. In fact, if you have identified them as marking the latter stage of a cold front going through, they actually herald the clear, cooler weather to come.

THE TALL TELL

In many games we're happy to reveal our true state of mind, especially when we're doing well—think of the smug new hotel owner placing that red plastic building on the Monopoly board. But in poker, experienced players try to hide a strong hand. They can only entice an opponent to bet against them if they hoodwink them into believing they have nothing special. Even the best poker players are human, and all humans struggle to conceal strong emotions. However hard we may try, strong feelings bubble up as involuntary movements, tics, and mannerisms. The poker pro knows the tics of regular players. The too-casual throw of a chip suggests this is a hand best sat out. The habits that reveal the truth behind the stoic facial expression are known as a "tell."

Nimbostratus, the rain blanket, promises hours of steady rain. We rarely get to see it against a blue sky, as it likes to arrive after a procession of other clouds. But if you happen to spot it approaching and then notice wispy white cirrus above the dark grey blanket, beware! Tall, unstable, potentially stormy clouds are hiding within the blanket. The nimbostratus promised monotonous rain and no violent weather, but the cirrus cloud you spotted over the top revealed its true hand. Very heavy rain and high winds are the least you can expect, and thunderstorms and hail are quite possible. This is the cirrus tall tell.

CIRRUS AND FRIENDS

The shape and patterns that we see in cirrus clouds are not complete forecasts, but they set the scene, offering a strong clue as to what to look for next. They are heralds. Mares' tails or jet-stream cirrus make us suspect that a low is approaching and encourage us to take a forensic interest in any new cloud types and wind changes over the coming hours. As we saw earlier, cirrus followed by cirrostratus means we can expect a warm front.

Before any new cloud type is seen, there will be trends in the cirrus itself. Like most other clouds, cirrus will cover more of the sky and will thicken if worse weather is on the way. They may lower, but because they inhabit their own band of the upper troposphere, their height is less conspicuous than their size and shape.

CONTRAILS

Passenger aircraft fly in the same high part of the troposphere as the cirrus clouds we see. The exhaust from their jet engines contains plenty of water vapor, but it also includes tiny particles that can act as the nuclei that help that vapor to condense. This is an ideal combination of ingredients for cloud formation.

Watch high jets pass over you and you will notice how sometimes they leave long, straight, white lines behind them and sometimes they don't. The long white lines are jet-made clouds known as contrails.

Contrail is short for "condensation trail," and the clue they offer us lies in the name. The aircraft are flying at a height where these clouds can always form, so the fact that sometimes they do and sometimes they don't reveals something about the moisture levels and wind in the atmosphere at that level.

The best way to think about contrails is not actually to ask whether they form, but how long they last. It is best to assume that contrails are forming behind every aircraft, then either disappearing instantly or lasting for long periods. The contrails that last a long time indicate moist air, and the ones that disappear instantly signify dry. A trend from a blue sky with no contrails toward lots of these white lines stretching over your head is a clear sign of rising moisture levels in the atmosphere. And we know that this is one of the signs of an approaching warm front—it's exactly the same principle as normal cirrus clouds filling the sky ahead of a warm front. The jet exhaust acts as a catalyst, a nudge.

If the conditions are perfect, these clouds can easily last half an hour. Contrails that not only survive for long periods but appear to grow, widening and spreading, are a sure sign of an atmosphere near saturation and a strong clue that wet weather is approaching. If you see these lines crossing the whole sky, keep an eye on any natural cirrus trends: There are likely to be some already. You can expect their numbers to grow significantly, nearly covering the sky, before the thin white veil of cirrostratus takes over. Rain is on the way.

To start with, most people see this as a binary clue: There are either contrails in the blue sky or there are not. But I'd encourage you to take a keener interest. Notice how there are often very short contrails behind aircraft and pause to gauge their length. Hold up an extended fist. If the contrails grow from one to two, then three fists, over the course of an hour, you are on the trail of a changing atmosphere, long before the brash white lines appear that anyone can spot.

Fat contrails might give you a day's warning before a front reaches you. The habit of noticing changing lengths will give you an increased sensitivity to air moisture that can stretch this warning to thirty-six hours.

Study the shapes of contrails: No two are the same. Early in its life, a contrail will often appear as a pair of white lines. This is not because of the plane's two engines but because the air flows off the aircraft's wings in a way that leaves a whirling vortex behind each tip. The vortices pull the contrails into two separate lines, which merge soon after.

Contrails begin as lines of great order and symmetry, drawn with white pencil and ruler on blue paper, but they soon lose their formality. There are great advantages to a cloud that starts as a straight line; our eyes find it easy to pick out the slightest deviation. Even as they hold their linear nature, contrails reflect every slight turbulence in the high air. The vortices that spin off the aircraft's wingtips can give the lines a ruffled appearance, like the crenellations of a castle wall.

Military aircraft designers do not like contrails. When they are testing the latest radar, thermal systems, satellites, stealth technology, and other wizardry to try to outwit the enemy, it's a bit frustrating to find that your billion-dollar jet has just drawn a fat white line across the sky.

DISTRAILS

Clouds are delicate. The conditions need to be right for them to form, but they must also stay right for them to survive. When a jet aircraft flies through any type of cloud, everything immediately around it is thrown into confusion. There is a whirl of fast, turbulent air and a big injection of heat from the engines. Most clouds shrug off this mayhem, but many high clouds are too delicate and fragile to do this, so they break. This will be seen in the sky as an anticontrail, better known as a distrail, a thin line of blue sky surrounded by cloud. Like contrails, distrails always form in the path of the aircraft that created them.

If the aircraft is cruising at a constant altitude, the distrails can stretch as far as the cloud it is disrupting. It is just as likely, though, that the aircraft is climbing through a layer of this cloud and melts a hole in it, sometimes known as a hole punch, like a burning cigarette through silk.

FALLSTREAK HOLES

Sometimes when a hole is punched through the cloud, the broken pieces come back together and form a new cloud. When an aircraft passes through clouds that are made of water, like altocumulus but near freezing, the aircraft can cause the clouds to freeze into ice crystals and then fall out of the cloud, leaving a hole. This creates a beautiful pattern: a blue-sky hole in a layer of cloud, with a feathery cirrus pattern inside it.

THE TWO MACKERELS

"Mackerel sky, mackerel sky, never long wet and never long dry." This weather lore is hedging its bets. Surely a sign should tell us if the weather is likely to improve or not. And, if not, has it any value?

We're going to get to the bottom of this mackerel-sky business, but there are two levels of engagement with this sign. First, there is a basic, nothing-to-fear, business-as-usual sign. And here it is: If you see a high cloud formation that could in any way be described as looking like a mackerel's skin, the weather is likely to change soon.

Then there's the seat-belt-on, know-what-to-look-for-and-what-it-means sign. This is all about recognizing that two totally separate cloud formations could be described as mackerel-like, depending on what a mackerel looks like to you. And since each one is a separate cloud type, each holds a different meaning.

"Mackerel sky" conjures up two different images in people's minds. Some imagine a series of nearly parallel solid lines, rippling zebra-like across the sky. Others see more of a stippled effect. Both of these are entirely accurate because there are different mackerel species with these skin patterns and more. As long as we learn to differentiate the two, we can read mackerel skies with confidence.

Cirrocumulus is a high cloud (*cirro-*) that is made up of lots of small puffs of convective cloud (*-cumulus*). The individual clouds are smaller than a pointed fingertip, and this is what gives the sky a stippled appearance. Like cirrus clouds, cirrocumulus sits well above the lower clouds and will not grab our attention; we need to look for it. And, like cirrus, it comes in many patterns, but the one we're interested in is wavy. In short, we're looking for a wavy, stippled blanket of high clouds. When this type of cirrocumulus is seen, it's a sign that a front is due in about twelve hours.

Before the bad weather arrives, cirrocumulus can form parts of stunning sunrises and sunsets.

We also sometimes see wavy patterns in much more solid-looking clouds. These are most likely altocumulus, forming alternating white and blue stripes that could also be described as mackerel-like. They look quite different from the cirrocumulus version. To tell the two apart, look at the size and form of the clouds. Altocumulus is more substantial: It is a bigger, more solid, and denser-looking cloud. The shadow test can help: If you see a wavy pattern in the clouds, ask yourself, Could the clouds cast real shadows? Put another way: Will the passage of these clouds in front of the sun cause it to switch on and off? If the answer is no, then cirrocumulus is more likely. If it's yes, this is probably altocumulus.

Waves in altocumulus mean that there is wind shear. This will always signal change, but it is as likely to be improvement as deterioration. This cloud on its own cannot tell you which way things will turn. To do that, we have to enlist the help of other clouds out there—are there any cirrus?—plus monitor the wind.

CLOUD STREETS

Mackerel skies are not the same as long, parallel lines of cumulus clouds. When cumulus clouds form a type of striped sky, the clouds are much lower, so they should not lead to confusion. If in doubt, bear in mind that cloud streets form parallel to the wind, while both of the mackerel-sky wavy patterns form perpendicular to the wind direction, like sea waves.

Cloud streets form when the rising warm air that creates the heaped cumulus clouds is partnered with cool, sinking air in a very regular way. This sets up distinct lines of cloud downwind of the warm areas, and clear, blue sky downwind of the cooler areas. Cloud streets are most likely to form when the shade cast by the clouds falls on the already cooler areas. This reinforces the difference between the cooler and warmer surfaces.

If you see cloud streets, consider what lies in the direction they're coming from, which will be the same as the wind direction at their level. The lines will point to either a coastline or some other interesting source of warming in that direction.

The space between cloud streets is usually between two and three times their height.

THE LENS CLOUD

I remember looking through buildings as if they didn't exist. We were in southern Spain, visiting family, and had decided to explore the coastal town of Tarifa. It's famous for its winds. Kite-surfers travel from around the world to enjoy them, and their colorful kites dotted the sky. I felt on my face the wind they were using, and we'll investigate its cause soon, but when I think of Tarifa, I think of a cloud. Walking along the beach, I turned to see a sculpted, curved, layered cloud above the sand-colored resort buildings. I recognized it instantly, like an old friend. At that moment I felt the buildings disappear and sensed the hills behind them.

Whenever a wind blows over high ground, it's given a bumpy ride. It can leap up and over summits, take detours around the sides, or funnel down through valleys. Whenever air rides up and over a mountain and then down the other side, it creates a roller-coaster effect in the wind, known as a mountain wave. The mountain wave is, of course, invisible, but whenever air is forced upward, the air cools, and cloud formation becomes more likely.

The tops of the mountain-wave roller-coaster are where the greatest cooling takes place and where clouds, known as lens clouds (*Altocumulus lenticularis*), are most likely to form. They often appear as lenses, but sometimes you'll see saucers and even flying saucers. Whatever we choose to call them, they're always unmistakable smooth, shallow domes. And they are a distinctive sign of the high ground that caused them. Lens clouds are a personal favorite for three reasons. They're easy to recognize, they offer a bold, clear sign, and they're beautiful.

Lens clouds form on the lee side of the summit that caused them. They are created by the wind but do not flow with it. This is because the wind that causes them, the mountain wave, is a "standing wave." The air is racing past the summit, but the shape it creates doesn't move. This is such a common feature in nature that it's worth a short detour to get to know it a little better.

Lens Clouds

Banner Cloud

Rotor Cloud

Mountain Wave Winds

If you watch a shallow stream as water flows over pebbles, you'll see lots of tiny waves just downstream of the stones. If the water flow remains constant, the small waves will hold their position and shape, but the water continues its journey downstream.

Imagine standing astride your small stream and focusing on one of these small stationary waves. Now take some red food dye and add some drops to the water, just upstream of your small wave. The red water flows up and over the pebbles, into your wave, and continues on downstream without stopping. The red color flows *through* the wave, constantly moving, even as the wave itself is stationary.

But if you dropped a large stone upstream of your pebble and wave and slowed the flow of water, you would very quickly see the wave break up. You have just disrupted the flow, and it's the flow that creates the standing wave. Why is this important when looking at lens clouds? Because the cloud indicates the wind wave and is just as sensitive to any change in the flow. It is impossible for there to be a significant weather change without a wind (flow) change, and this will mean the lens cloud must also change or disappear.

A line of these clouds may reflect a series of summits but doesn't always. A single summit can create an undulating wind all on its own, which can lead to a cloud at the top of each crest.

It is fairly common to see a vertical stack of these shallow domes, but obviously that cannot mean mountains piled on top of each other. A stack of lens clouds is a sign that there are alternating layers of moist and dry air in the atmosphere. The moist air creates the first lens cloud, the dry layer above it leaves a gap, and so on; half a dozen stacked saucers is rare but possible. There is a tidy French term for this effect: *pile d'assiettes*—a pile of plates.

Lens clouds form only if the atmosphere is stable and the wind is fairly strong and constant. It's likely to be windy on mountains, but you can expect fair weather. Light snow or rain may fall from the biggest of these clouds, but the lens clouds will change markedly or disappear long before any seriously bad weather can arrive.

When the conditions are right, lens clouds can be seen above mountains all over the world.

ROTOR CLOUDS

Whenever a fluid passes an obstacle, there is turbulence. It can be seen when a twig dips into a stream, steam passes a kettle spout, or clouds form downwind of mountains. We saw the same turbulence in the formation of the rebel dune in snow in chapter 9.

A rotor cloud forms when turbulence leads some winds to blow back toward the mountain. As the name suggests, this cloud is the visible part of a rotating vortex of wind on the lee side of the high ground. It can lead to a surprising array of shapes and forms, often a warped cumulus, depending on which parts of the rotating wind are cooled enough to cause condensation.

If you see a peculiarly shaped cloud in a mountainous region, look from the cloud in the direction the winds are blowing and you should be able to spot the summit that created the rotating wind. Pilots are expected to spot these clouds: They are a sign of winds that are dangerous for aircraft.

THE BANNER CLOUD

In mountainous country you may also come across a cloud that streams away from only one side of a mountain summit: the banner cloud. The first time I saw one was in the Alps. My brain decided that I wasn't looking at a summit and a cloud at all: It was a rock steam train.

When moist air is blown into the summit of a mountain, it hits a wall. The air can't immediately escape and is temporarily stopped in its tracks, but air is still arriving with the wind. This leads to compression and higher air pressure, and whenever air is compressed, it warms.

This must also mean that there is an area just downwind of the summit where the wind is flowing away but not arriving quickly enough to replace it. It causes lower pressure and expansion; the air cools. Now we have slightly warmer air on the upwind side of the summit and cooler air on the downwind side. A cloud is much more likely to form on the cooler downwind side. The wind continues to pull at this cloud, giving it a streaming shape: hence its name, the banner cloud.

This cloud grows more likely the more impressive the summit. There have been famous examples at the Matterhorn in Switzerland and, unsurprisingly, on Mount Everest. There are no banner clouds downwind of rolling hills.

This cloud shows us that the air is near saturation at that level, but its main job is to fly a flag and display the wind direction for us. It is a simple cloud with a simple meaning.

CLOUDS FLOATING ON SOUP

The lowest part of the atmosphere is the only area strongly influenced by the sun's radiative heating of the Earth's surface. And the part that touches the ground is most strongly affected. It behaves differently from the higher parts. The low layer is known as the planetary boundary layer and it's a region that experiences massive daily fluctuations in temperature, moisture, and local winds.

If we think of a pot of soup being placed on a gas burner, before anyone stirs it there are turbulent changes in the soup at the very bottom of the pot: The temperature shoots up, steam bubbles begin to form, and there is turbulence, even as the top of the soup stays flat and calm. Every day, the sun's heating of the Earth's surface acts as the gas burner, and the lowest layer of the soup is churned, but there is little stirring in the higher layers of the troposphere. The lowest layer also experiences the most extreme cooling as the land loses its heat at night, through radiation. To stretch the metaphor, each evening the pot is taken off the gas and put on ice.

Why do we need to know this? Each day and night a distinct layer close to the ground follows its own rules. The temperature, humidity, and winds will be different in the lowest level from those a bit higher. The layer gets thicker as the day progresses and shrinks again at the end. It might shrink to 325 feet (about 100 m) after a cold night and be 1.25 miles (2 km) thick after a hot day. The warmer the region, the thicker this layer: It may be a mile and a quarter in the tropics but is more likely to be nearer 160 feet (50 m) in the Arctic. The boundary is not visible, but you may spot it by looking for certain features. The base of low clouds normally marks the top of this layer. If you notice low cumulus clouds with rounded tops and flat bases moving across the sky, the flat bottom is sketching the divide for you.

At the boundary layer, there is often a temperature gradient change and the air gets warmer. As we have seen, warm air on top of cold air equals stability, and in this case creates an effect known as a capping inversion. This leads to smoke, dust, and moisture becoming trapped in the bottom layer. If you have ever looked down from high ground to see a distinct layer of dusty, smoky, or misty air coating the land, you are peering into the planetary boundary layer. Drop into this layer and, in weather terms, it will feel like another country. Capping inversions are most common at the start of the day, when the heat has radiated out of the land and the lowest layer of air touching the ground has cooled sharply.

It's worth getting into the habit of spotting the level of the bottom of low clouds or the top of misty, smoky, or dusty air when you get the opportunity and noting that this is the top of the soup and that the weather above and below will be behaving differently. Looking up from within this layer at the lowest clouds, you'll also notice that they usually move in a similar but slightly different direction from the wind you can feel closer to the ground.

THE HEAPED BLANKET—STRATOCUMULUS

The last cloud we're going to meet in this chapter is stratocumulus. It's the most common cloud in the sky, so why didn't we meet it before? Because it's a bit dull, and a fairly bland sign, too.

Stratocumulus is a low cloud made of big white and grey lumps. As the name hints, it is a blanket cloud consisting of cumulus parts. It forms when either a stratus blanket breaks up or cumulus clouds bubble up in good numbers and then meet a very stable layer. The rising clouds are stopped on their upward journey at the same altitude, often at a layer of much drier air or a temperature inversion at that level. They regularly cover more than half the sky and are very common over oceans.

They are slightly annoying clouds, refusing to neatly fit either blanket or cumulus models. They are very common, but mainly because the name is a broad label that can be applied to a variety of similar forms. There is clearly unstable air in the lowest band of the atmosphere, but this is capped by very stable air. The sign is simple: The blanket indicates a settled atmosphere and settled weather to come, with no change likely in the next twelve hours. Precipitation is unlikely, and the winds will be modest.

The way to tame these clouds is to remember that a blanket made of anything, even balls, is still a blanket. And blankets mean no sudden change.

All blanket clouds have a noticeable effect on temperature. Like wool blankets, they keep the warmth in at night and in winter, but they can also block out the heat during a summer day. Stratocumulus are softening clouds. They take the edge off things: Frosts or heatwaves are unlikely when blanket clouds are around.

If you are ever asked what the weather is doing by somebody standing under a sky covered with stratocumulus, the answer to the question is "Not much."

CHAPTER 12

Thirsting for Signs: An Interlude

IN MAY 2019 I had been sheltering from determined gusts and face-biting sand. Nestled between dark rocks in the Sharjah Desert in the United Arab Emirates, I peered down a steep rock slope. A hundred feet below me Abdul, a friend of a friend, was sensibly sitting out the sand squall in the air-conditioned Land Cruiser. I pushed myself harder into the rock crevice, not to shelter from the wind but to hide from Abdul.

Once I was happy that he couldn't see me, I took out a flask and drank mighty gulps of water. It was Ramadan, and Abdul would take neither food nor water between dawn and dusk. In empathy, I was happy to go without food in the daylight hours while we were together—in the heat of the desert it feels surprisingly natural—but water was another matter. That I could not forgo.

When observing Ramadan, it is a good idea to take plenty of rest during the day; without food or water, exertion is inadvisable. As a favor, Abdul had been kind enough to offer to drive me to the places I wanted to get to, but to conserve his body's water he needed to be economical with effort and avoid the heat. He didn't say this or come close to complaint, but it was obvious and sensible.

My business on this small foray into the desert was impossible without some exertion. A rhythm developed: I would spot something requiring investigation from our air-conditioned comfort and ask Abdul to stop the Land Cruiser. I would then step out into the ferocious heat and venture into the distance. Once out of sight, I swigged from a water flask, then pulled out my notebook and got to work.

On this occasion I had spotted some extraordinary patterns in the rocks. Dark stone was laced with white veins. This is common enough, but there was a brilliance and vibrancy to the white veins and their marquetry that shone across the desert and pulled me toward them. Up close they were more remarkable still. It was a Jackson Pollock painting before his first drink of the day—perfect order dressed as randomness. And that is always a thing of beauty. I took photos, made a sketch, drank some more water, and waited for a lull in the wind.

Pollock may be a worthwhile analogy. Most people on first viewing a Pollock painting see no structure or meaning and are dismissive. I was no different. I'm still skeptical of the loftier claims, but my opinion changed when I heard that his paintings are fractal. Take any part of a Pollock painting, large or small, zoom in, and the same patterns appear before your eyes. Even though I once carved a path through the irritable energy of a crowd in the Museum of Modern Art in New York to peer ever closer at one of his paintings, I cannot claim to have verified or quashed this theory. But, if true, it suggests not only order and meaning in the work, but that this is threaded throughout and deeply imbued. And we find this so often in nature. A tree in the wind may appear at first as an indeterminate shape—green noise—but closer inspection reveals order, first in the canopy shape, then the branch patterns, the twigs, the leaves, their serrated edges, veins, the colors, and even the sounds.

I felt ecstatic about the patterns in the rocks and determined to decipher their meaning. At the same time, I felt guilty about Abdul below. This is a conflict that I have often felt, and it can make me uncomfortable: Why should someone suffer greater thirst just because I'm curious about such things?

Then, over the mountains to the east in the Emirate of Fujairah, I saw a very different pattern. This one was in the sky. It was the sign I had hoped for but not expected.

I scrambled down the rocks, back onto the dunes, and made the inelegant run of the sand-crosser to the car. I squeezed in, trying not to let the whirling sand follow me, and turned to Abdul.

"OK, it's time. Let's go."

"Where?"

"To find some rain!"

It rains on only five days a year in that part of the world, and almost never in May. But I felt confident. Some signs are hard to resist.

There had been rumors of showers in the days before, when I had been working in Dubai. I had peered through the city's tinted windows and seen clouds growing tall over the coast, their tops bending toward the land, which tallied with reports of moist air masses coming in off the ocean. I asked around to see if anyone knew where it had been raining, but my inquiries were met with glazed indifference. I had my theories, but local knowledge might make the difference between a fruitless search and the jackpot. I wanted to speak to someone who still knew the land and the sky.

A question I am asked regularly: How do you get in touch with indigenous wisdom? Many people love the idea but are bemused by the challenge of finding something that does not sit in any directory, even on the internet. It is always a process of stepping stones. The "six degrees of separation" concept argues that we are all connected through six steps of a social network—meaning people who actually know each other, not people who "like" one another online. The rule in any region for meeting people who still know the old ways is similar.

Maybe the first person you speak to doesn't know what you're after or may not understand the nature of the question. But they can introduce you to someone who can make sense of what you're asking. That person doesn't have the answer, but knows a person who might. They don't, but they know someone who probably does, and so on. Through this pedantic but effective process I had set up a meeting in the desert with a Bedouin elder named Mohammed. The plan was nebulous. I would head out into the desert, try to find the rain, and then, whether I succeeded or failed, the day would end with Mohammed. Abdul had kindly agreed to help me with this sketchy plan.

The thermometer in the Land Cruiser gave the outside air temperature as 95°F (35°C), but it was still early as we left the city behind us and watched the mountains growing ahead. Abdul explained that the temperature differences between the coast and inland were stark, but no less than the difference that altitude brought. Up on the Jebel Hafeet—the "empty mountain"—on the border with Oman, it might be 45°F (25°C) cooler than it was lower down. The mountain he referred to was little more than four thousand feet high; a temperature difference of that range can be explained by the air near the low ground being heated by land that the sun has cooked.

In the few desert places where there is water, the effect on the microclimate is extraordinary. The water is cooler than the sand, so the air above and around it is cooler, too, but the water allows plants to grow and they are cooler than the sand. The water, greenery, and cooler air allow an influx of animals. A little water in the desert has a cascading effect on temperatures and life. Two lakes had been built on the desert side of the built-up area of

Dubai city, and locals flocked to the cooler air around them. In typical restrained Dubai fashion, they had been sculpted in the shape of hearts and named Love Lakes.

We sped past the camel hospital and racetrack. Abdul explained that the fastest camels can run at 40 mph (65 km/h) and change hands for more than a million dollars. I'm not aware of any cultures that don't bet on fast or strong animals one way or another, and camels are a better bet in this part of the world than horses. About ten years ago, in the Emirate of Ajman, I visited a desert training facility for racehorses, complete with a horse swimming pool, quite surreal but very modern Arabian.

Soon the road led us into the desert. There was a line of bushes on each side of the road and the sand changed color on either side of them. Abdul confirmed my suspicions: The bushes were planted deliberately and irrigated year round to act as windbreaks. In high latitudes, where roads are vulnerable to snowdrift, they erect windbreaks. In places like Canada the positioning is an art and a science: Put the windbreak in the wrong spot and you exacerbate the problem by creating a wind shadow in the wrong place, causing the snow to build rebel dunes on the road, among other types of accumulation. I was delighted to see an example of the art form in a desert and stole glimpses of rebel dunes through gaps in the greenery.

The roadside bushes disrupted the airflow—that was their job—and this caused the change in sand colors. The heavier, redder sand had dropped on the windward side of the bushes, and sand that was lighter, in both color and weight, on the downwind. This is a trend you will find all over the world wherever wind meets sand, not just in deserts. I have seen it on either side of roads and rocks in Africa, Asia, America, Australia, and at my local beach in Sussex. The lighter-colored sand is lighter in weight, too, and drops on the downwind side of an obstruction. With sand, light is light.

The temperature increased to 97°F (36°C) as Abdul explained that it was a full moon in the Ramadan calendar, which I found mystifying. The moon had been full two days before, but I didn't quibble. The man I sat next to was heading into the desert for my benefit without food or water. Full moon could be whenever he said it was. We slowed as we passed through a village to allow a gaunt figure to cross the road. The thin man held a palm frond that I thought was for use as a fan or for shade, but then he reached up to touch the trees with it. Abdul explained he was fertilizing the palms that lined the road.

At my request, Abdul pulled over to give me a chance to explore a small plateau. I walked away from the vehicle, over the rocky land, and looked all around. The visibility was still good, which allowed a view over great distances to the emirates of El Ain and Sharjah, as well as to Oman in the southeast. But that meant the air was still fairly dry, which did not improve my chances of finding rain.

The sun had climbed and was now intense—the black cover of my notebook was too hot to hold. I planted a stick in the ground and marked the end of the shadow. It's an old habit and a rewarding one: It always yields something beyond the thin shadow. By stopping to look closely at the ground, my eyes had time to adjust; the colors of the land were sifting from a bleached brightness into varying shades. And then I saw a flower. I followed its green tentacles and suddenly I was standing over a small carpet of green veins and small yellow blooms. I grabbed a sprig and ran excitedly back to the vehicle, eager to get Abdul's take and save him the exertion.

"It has rained," he said.

"Really? When?" I was barely able to contain my excitement. Abdul strolled to a nearby carpet of the same flowers that I had missed and lowered his head to look more closely. "In the last couple of days."

By now I was almost skipping on the spot—and all because Abdul had done some skipping of his own. He had skipped identifying the flower and gone straight to the sign, which is such a rare treat. It is much more common for me to spot something anomalous, asymmetric, or otherwise intriguing, having to pursue its meaning via identification, then further questions or later research. I contained my joy and led the investigation in the opposite direction to normal. "What flower is it? Do you know its name?"

"It has many names. I know it as the Bindii flower."

Later I was able to confirm it as *Tribulus terrestris*, well known in many dry parts of the world for flowering with the first rains after a dry season. Perhaps the flowers planted the thought, and I don't think I imagined it, but the air now felt more humid than it had at the start of the day.

We got back in the car and drove on. We were now into desert proper and well past any attempts to corral the sand. There was a red warning sign, written in Arabic and English: Mobile Sand Dunes on Road Area. We refueled in a village that bordered an oasis. There were trees, and the sounds of birds that we had not heard for a couple of hours. Even the village shops proclaimed water, life, and its fruits, one selling honey and bees. The shadows were very short now: 98°F (37°C).

At the next requested stop I climbed to gain height and perspective, and to my delight, the views were poor in all directions. The sky was now filled with a generous assortment of cumulus clouds; the air felt muggy and looked heavy with humidity and dust. The mountains to the north were faded, lacking the brilliance we'd seen earlier in the day. I noted that the ridges ran north-northwest to south-southeast, and the dip, the angle at which the rocks lay, was down from northwest to southeast. I marked another shadow and took some photos of sand tail patterns. Prompted by my interest in the rocks, Abdul explained that regular earthquakes were felt in the local rocky areas but not in the surrounding ones, where the vibrations were soaked up by the sand.

We drove across the desert toward higher, rockier ground, then down again into a dune range where we stopped to deflate the tires to give us better traction over deep sand. I continued into the dunes on foot for almost an hour, burning through water fast, then circled back, using a distinctive rock feature as a convenient landmark. I felt myself swinging around this visual anchor. A rich plethora of sand compass patterns was arrayed before me, each shaped by the wind on its own private scale. The full fifty-foot dunes had waves impressed on them, ripples over those, and sand tails around the scattered small rocks. Sadly and predictably, a pale sand tail stretched away from the neck of a plastic bottle.

I saw a familiar and strange neatness in some trees and realized it must have been the camel browse line: The base of the trees had been shorn neatly at the line that the camels could reach. Before cresting the final dune, I hid my water bottles. Back at the car, I asked Abdul about the isolated clump of trees I had found.

"These are the desert cedars. Their sap is poisonous, but when mixed with sand it is used as an antiseptic for camels with cuts."

I got back into the Land Cruiser: 102°F (39°C). We drove on, and then I spotted the black rocks with the white veins.

The distant cumulus clouds bubbled up energetically, changing form in seconds. The ones I was looking at were the parents of the gusts of sandy wind that had me sheltering among the patterned rocks. In the distance I could see how this chain of events had unfolded.

The warm, moist air had come in off the ocean and was passing over the desert plains. As long as the land remained low and flat, the only effect it had was to create a slightly muggy feel and reduce visibility.

By early afternoon the sun had heated the ground sufficiently to create some local convection and instability, which had led to the clumps of cumulus that were scattered over the desert. If all these factors remained constant, the weather would, too, but the conditions were ripe for something more theatrical. The air was full of energy and ready to display it. A little local heating would help modest cumulus clouds to form, but for these to grow to towering rain clouds it would take a bigger trigger. And that trigger was the mountains.

The cumulus clouds grow taller than wide, signaling moist air, thermals, and instability.

The mountains forced the humid, unstable air upward. It expanded and cooled, the water vapor condensed, and more heat was released. The modest heat from the sunbaked sand had started some minor convection before allowing things to settle back to equilibrium—fluffy cumulus. But the high ground changed everything. Every bomb needs a detonator, a smaller explosion that is just enough to get the chain reaction going properly, and the ramp of the mountainside was that detonator. I used the sun to get a bearing on the clouds and the mountain.

"Can we get here?" I held the map across the hood of the car, ran a finger along the line of my bearing, and laid it on a point on the windward side of the jebel.

"Yes."

Pausing to reinflate the tires at the edge of the dune system, we then sped toward the spot where I estimated the maritime air would hit the mountainside. As we climbed into the mountains, we crossed into the southern part of Ras Al Khaimah, a different emirate, but the political lines were of no interest to me as long as they didn't stop us.

Soon we were about a thousand feet above sea level and there were plenty of trees on either side of the road, which was a good sign. It meant water reached here at times, whether straight from the sky or flowing from higher up. I couldn't tell from the car. Then, very quickly, the sky changed all around us. I asked Abdul to pull over, and I wandered down from the roadway into the rocks. There were mountains on either side of us and I wanted to get a better view of the sky.

I felt a sharp gust of cool air, the first coolness I'd experienced outdoors in days. Looking up from the bed of a broad, gentle gully, I could see that the clouds were now morphing too fast for me to keep track of them, and light levels were dropping. I heard dry leaves rustling among the rocks and watched one leap into the air and whirl upward. I felt the first raindrop on the back of my hand as I made notes, then took off my hat, looked straight up, and felt the next drops fall on my face. A minute later there was a downpour and thunder rolled down from the mountain. The clouds were moving in a totally different direction from the wind lower down, which whipped up and backed suddenly. Water ran off my nose as I stashed the notebook in my knapsack. I shivered as a cold trickle worked its way down my back.

I celebrated, waving my hat at the sky, then ran back to the road and slightly higher ground, but not too high. Summits are risky when there's lightning nearby, but I don't like being in gullies in dry countries when rain is in the vicinity, either: There are too many horror stories. In deserts, the rare storm rains are channeled along wadis, ravines, and gullies, and water levels shoot up in seconds. Locals are brought up with stories about the dangers of flash floods and know to be careful, but travelers never imagine that their lives will be snatched by water in the desert. Isabelle Eberhardt, the Swiss explorer and seasoned desert traveler, died in a flash flood in the Algerian Sahara in 1904.

The trees I passed on my way back to the car weren't surviving on a few drops: They were being quenched by deep roots that drank from the soaking of flash floods. Back at the Land Cruiser, the thermometer revealed that the air temperature had dropped 14 degrees to 88°F (31°C).

We drove east and watched the road turn from wet jet-black to dry every couple of miles across the mountain range, reflecting the isolated downpours that had been dumped by the runaway unstable air. Near the coast, I could watch the humble birth of the storm clouds in the harmless-looking cumulus that were still sprouting.

That evening I met Mohammed as planned near a settlement in the desert. We took a mat from the back of Abdul's car and laid it on the sand to sit on. Abdul drove off, tires wrestling the soft sand, after promising to return in a couple of hours. The sun had set, the fast was over for the day, and we drank coffee from a Thermos, ate dates, and talked as the last of the light fell from the sky.

I thanked Mohammed and explained why I was so happy he could meet me. Before I had finished my sentence, Mohammed interrupted to tell me I would need twice the number of camels I thought I would if I planned to cross the Empty Quarter. "If you think you need ten camels, take twenty. Some will die." His hands danced in front of a perfectly still body. "And eat the desert fish. It is a lizard. It swims in the sand."

It took a few attempts before I was able to explain that my aims were more modest. And I did this in stages, first by pointing to stars and explaining my names for them and what they meant to me as a navigator. Mohammed picked up on this, his expression changed, and he explained his names for the same stars, how they were used for direction by the Bedouin. The methods he mentioned were familiar—they are shared the world over. It is the names that change. Then I led the conversation toward weather lore. I used "Red sky at night, shepherds' delight" as an example. I didn't think it would resonate with him in terms of weather signs, but I thought the tying of lore to shepherds, weather, and wisdom might work. To my delight, it did.

"This is right question!" His finger was wagging. "When the wind blows four days from the south, do not sleep in the wadi. Heavy rains will come!"

This hit the mark, and Mohammed must have seen the excitement in my face. He began to elaborate, explaining that a southerly wind for a few hours means little, but if it lasts into the fourth day, a deluge will follow. This would be dangerous to anyone in the wadis, the riverbeds that run dry for months before storms drive torrents along them. He went further, but he lost me as he tried to explain how this sign was tied to the Bedouin belief that you trust no one whose father you don't know. Perhaps he was reminding me that the wind is the father of the weather.

I encouraged him to continue, and he switched to talk of a northerly wind. Mohammed's English was fairly good, with a strong accent, but on the few occasions I lost the thread it was usually because of an Arabic name for a wind that was new to me. We would then point to the stars for a direction label, and they served admirably as our interpreter. The northerly wind signaled the coming of winter, but it was also a sign to wear a mask. Mohammed explained that the northerly wind indicated that disease would soon arrive.*

The easterly wind, the one I had felt in recent days, was the one that brought clouds and sometimes rain. It was wonderful to hear Mohammed explain how the Bedouin tradition was to associate days of easterly winds with short rain, not long like the rains the southerly winds bring. The science behind these observations was easy to see: Easterlies brought warm, moist air and short, sharp convective showers; the southerlies indicated a seasonal shift in air masses and subsequent frontal rain.

"The mountains, they are magnetic for the rain. This is why they are green in places. The green is where the rain falls." Mohammed had hit his stride now, and I adored his poetic turns of phrase. Who needs relief rain when you have magnetic mountains?

I asked about animals and weather, and he told me of a spider—"not spider, like spider family"—that turned from black to red when rain was on its way. I did my best to transcribe its name, *khumfasir.* I tried to draw him out, to glean more details, but I don't think he knew the animal personally and have since failed to find any other reference to it or to identify it. Another search that continues.

---

* This was months before the first stories of COVID-19 hit the news, but I've wondered many times since if we might soon be copying this tradition: The cold winter winds encourage us to wear masks before politicians or scientists mandate it.

The coffee and dates on empty stomachs gave us both energy. The work for the day was done, and the conversation ranged more widely. We both laughed a little. Mohammed told me that camel's milk was excellent for male sexual performance.

The lights of the Land Cruiser appeared over the crest of a dune. It had been a fortuitous meeting and peculiar in the way that it was back-to-front. Normally, local knowledge helps me find what I'm looking for. On this occasion I had been so lucky during the day that much of what Mohammed told me reflected my earlier experiences. Both parts, day and night, encountering the signs and then the wisdom, were fortunate, and all travelers must be happy to accept any luck in the order it chooses to come their way.

I thanked Mohammed for his time and for sharing his wide-ranging wisdom. Then, before they headed off and left me on my own, I asked Abdul if he could meet me in the same spot in the morning. When they were out of sight, I pulled some bread from my pack and attacked it. It was hard to sit still. I walked off the coffee under the stars for a couple of hours, then lay down on the mat and watched small clouds shrinking until I fell asleep.

CHAPTER 13

The Local Winds

The Gap Wind • The Channel Wind • The Split Wind •
The Rebel Wind • Summit Winds • The Mountain-Wave Wind •
The Sea Breeze • The Land Breeze • Mountain, Valley, and Lake Breezes •
Joining Forces • Winds with Names • The Crosswinds Rule

I N THIS CHAPTER we'll get to know the local wind characters to look for on our journeys. There are two main families: The winds shaped by the land, and those created by heat differences.

The characters in this chapter are the archetypes—each one explains a core principle—and through them we can understand almost every other minor character we will encounter, too. This concept is so fundamental to understanding wind that I'd like to underline it by taking a moment to look at a parallel area and use that to explain how it came to underpin my thinking and approach in this area.

For decades I'd been fascinated by the way in which Pacific Island navigators had learned to find their way between small islands and across hundreds of miles of ocean. They used many methods, but one in particular is relevant to us. When water waves bump into an island they bounce back, bend, and fan out: They reflect, refract, and diffract. They do this according to the laws of nature, creating dependable patterns in the water. The island navigators sensed the anomalous patterns in the waves through the motion of their canoes: The patterns indicated that land was nearby and the direction to head to find it.

The thing that blew my mind, and led me to write *How to Read Water*, was spotting these patterns in a pond near my home in Sussex. I suddenly realized that the laws applied globally and over very different scales from what I had imagined. The patterns were not confined to the Pacific or even to oceans, but could be seen as small ripples in ponds all around the world.

This insight drove the book about water, but it also showed me how to approach some of the concepts in this one, and proved most powerful when I was trying to read winds. The patterns are different from those in water, but again, the laws that underpin them are global and apply over wildly different scales: As soon as we recognize them we can look for them in every land and on every scale. The Levanter is a famous wind in the western Mediterranean (related to the levant wind that we met earlier). It blows over thousands of miles and affects millions of people but, as we'll see, is a close relative of the breeze we feel between two buildings.

Every landscape has its own relationship with the wind. The landscape shapes the winds but is also shaped by them. When the land influences the wind, it does it through the physical shape of its forms, the topography, and through its unique heat fingerprint.

The first set of characters we shall meet are the topographical winds, local winds born when the regional main wind, driven by the pressure systems, encounters certain land forms.

For the wind to reveal all that it can, we need to hold on to three simple truths:

1. *The air is never perfectly still.*

2. *There are reasons for what it is doing.*

3. *We can sense and understand these causes.*

If we keep them close, we will start to find meaning in the smallest wind signs.

THE GAP WIND

In 2007 I set sail single-handedly across the Atlantic. I don't write or even speak about this journey very often, for several reasons, principally because it mostly went according to plan. The main challenges were of logistics before setting sail, then heat, boredom, and loneliness during the voyage, none of which are essential ingredients for great storytelling. If the whole thing had been a disaster but I'd survived, I'd probably have found the raw material for a longer account. As it was, there were, to my mind, only three poignant aspects to this journey, and they are quite personal.

First, it was the culmination of many years of study, training, preparation, and effort, and it marked an important milestone in my goal of learning to master conventional navigation, which then allowed me to focus more fully on natural navigation.

The second is that it was an extraordinary way to do this journey. You will hear of single-handed transatlantic sails—they are not the rarest journeys undertaken, but this was quite extraordinary in its truly solitary nature. Most long-distance single-handed sailing is part of a race or other organized event. It is rare that one person gets into a boat and sets sail across an ocean on their own with no organizational support or framework. When I sailed out of the marina in Lanzarote, the Canary Islands, I don't think there was a human being, including family, within a thousand miles of me who knew what I was doing. And that's pretty much the way it remained until I approached the Caribbean twenty-six days later.

The third interesting aspect was the slightly hellish first forty-eight hours. I had chosen to depart from Lanzarote for a forgivably stupid but very human reason: I was familiar with it. I had walked and sailed around the island in the past, and my logic was the fewer unknowns at the start of this undertaking the better. The only problem was that Lanzarote lies at the northeastern edge of the archipelago and I was heading southwest. I had to thread my way through the island chain before reaching the open waters of the Atlantic.

There are numerous hazards for any boat doing this, but single-handed, it was testing. I saw a lot of other sailing boats, but the commercial vessels—not least the almost aggressively oblivious fishing boats—were more worrying. The most unnerving problem of all, though, was the "acceleration zones." As the winds arrived at the Canaries, they were squeezed between the sharp volcanic peaks of the islands and metamorphosed. They changed in an instant from benign breezes into violent gusting winds of force seven or eight —enough to throw a yacht with full sails into a dangerous dance. The suddenness of the change was physically and mentally exhausting. It meant constantly studying the surface of the water for the footprints of the wind, ruffled patches that might offer a few minutes' warning. This was challenging by day and near impossible at night. My interest in "gap winds" is now etched into my psyche.

Whenever a wind is forced through a narrow gap it accelerates. This is true of all fluids, gas and liquid (it's why we put a thumb over the hose opening to make the water go farther). It is also why, if you see a river pass on either side of an island, you will notice that the water is faster in the two narrow streams than when they rejoin.

The acceleration of winds through gaps has been noted by humans for thousands of years. Theophrastus, the ancient Greek scholar we met earlier, noted 2,300 years ago the acceleration of the wind and the parallels with water: "For a wind passing through a gap is always more forceful and vigorous like a current of water. Concentrated, it has more thrust. That is why, when there is a calm elsewhere, wind is always present in narrow passages."

In that last sentence, Theophrastus has also picked out the interesting way in which gap winds can make us feel like a wind has materialized from nothing. If a breeze is very light and we are less sensitive to it than we might be, it is often a gap wind that makes us notice the wind for the first time.

We will all have noted to a walking partner that "the wind has picked up," and in a sense we're right: It has indeed accelerated. But if we also mention that "the wind has died down again," the truth may be that the main wind has been a light constant breeze, but we have passed some gap upwind without spotting the hand it has played.

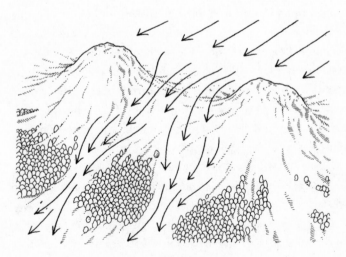

Gap Winds

The mistral is a famous gap wind, acting on a grand scale as it blows from the north and over southern France toward the Mediterranean. The wind is set in motion, like all winds, by a difference in air pressure. In this case it is often a high-pressure system over the Bay of Biscay to the northwest and a low-pressure one in the Mediterranean to the south. The winds then have to navigate the Alps and Cévennes mountains. The hills in these ranges act as thumbs over the hose, and the wind accelerates to infamous speeds in the valleys. The ancient geographer Strabo tells us that it hurled men from their chariots. The wind has changed little since Strabo's time, and more recently the writer Nick Hunt traveled to the region to meet this wind and was not disappointed with his reception:

> It was like being under attack. I crossed the street in defensive posture, shielding my eyes with a chilled hand . . . the tricolours on municipal buildings were doing their best to tear away from their flagpoles, and people walked at forty-five-degree stoops, their hairstyles heading south.

Like most gap winds, the mistral is most common in winter, partly because seasonal factors are driving the pressure systems, but also because gap winds are most extreme when the air is cold. Cold air is denser and can't rise over high ground the way warmer air can. The colder the air, the more gravity pins it to the ground and the more it is forced through the gap.

The gap wind will blow over short distances or long. If a landform narrows steadily to a gap, the wind speed will build steadily and climax at the narrowest point. The Santa Ana winds in California vary from light breezes near their birth at the high-pressure end to more than 60 mph (97 km/h) when they are pinched between the coastal mountains of the San Gabriel and San Bernardino ranges.

The gap can be a lower point in high ground or a dip in a plain. Depending on our perspective, the gaps might be called passes, cols, saddles, notches, gorges, or any alternative from a long list of terms. The key point: If there is a narrow lower route for the wind to follow across a landscape, it will accelerate.

The gap wind is too well known to be overlooked on the grandest scale—local residents know its habits in places like the Strait of Gibraltar—but as the scale comes down it will try to pass unnoticed. We need to keep in mind that winds will accelerate between two high points in a landscape, whether they are great summits, two woods, or even a pair of trees.

The wind we met earlier, rushing past the Spanish town of Tarifa, was a gap wind. Tarifa is just up the road from Gibraltar and all the winds that pass it are squeezed between the land of Europe and Africa.

THE CHANNEL WIND

Winds that began as gap winds often continue following the low route and become the winds we feel blasting along valleys a long way from the gap. Once a wind, especially a cold one, sinks into any valley that runs roughly in the same direction as the pressure gradient that is taking it, the wind will follow the valley. Valleys corral a wind and bend it along their axis: They channel it.

I'll never forget a few days spent exploring the Yorkshire Dales, a series of valleys that tend to run west to east off the north-to-south backbone of the Pennine Hills. It was mostly overcast, and I was using the wind-sculpted trees as my main compass. After drifting off course early on the first day, I realized that I needed to recalibrate my instruments. Instead of the treetops following the neat, dependable national trend of being bent from southwest toward the northeast, all the trees in these valleys were bent from west to east. The valleys were aligned west–east, which meant the winds were, and so were the treetops.

Where valleys reach the coast and there is an offshore wind, the channel effect means there will be bands of much stronger winds quite far out to sea and calmer waters nearby. The winds fan out gently in the bands, and the patterns they leave in the water are easy to spot, especially if viewed from high ground and if the sea is not too rough. Look for darker, broadening bands that reach out over the sea from any low points along the coast. The effect in places with undulating cliffs can be beautiful, and it's thrilling to see the shape of the land painted on the sea.

THE SPLIT WIND

Many ancient authors were enthusiastic students of nature and wove their imaginative accounts over a well-observed reality. Homer's works are rich with fantastic creatures and events set against a backdrop of sea, land, and sky that behave in a way scientists would recognize today. Apollonius of Rhodes, an Ancient Greek poet, took time away from describing Jason's Argonauts to recount that "Opposite Helike the Bear there is a foreland called Karambis, steep on every side, and presenting to the sea a lofty pinnacle, which splits the windstream from the north in two."

When winds meet islands, they accelerate in some places, but their direction changes, too. If they cannot go under or through the land, they must go over or around it. This is obvious when we think of islands jutting up from the flat sea, but we can experience it in everyday landscapes, too.

On one of my most regular local walking routes there is a natural stopping point. At the edge of fields used for sheep pasture, there is a raised grass shoulder, and I like to sit there in good weather, perhaps for an hour in summer or less in winter. At this spot, I can always tell if the day's walk was planned or the result of restless spontaneity: There will be a flask of tea and something to eat in my pack if any thought has gone into it.

There are good views in most directions, almost as far as the sea to the south. To the southwest I can see a small hill I know well, having walked over it from every angle hundreds of times. It has taught me much and has appeared in most of my books, always anonymously, as it has no recognized name. There must be thousands of such hills in England, dominant enough in their own small neighborhood, yet denied a title.

My normal route takes me on a circuit and leads me past this spot on the northeast side of the hill. The shape of the land and the southwestern character of our prevailing winds mean that I can usually expect to meet two winds on this stretch. The hill splits the southwesterly wind and I feel both, the southerly to begin with and the westerly soon after.

Even on the many days when we have a southwesterly, the wind is rarely ever perfectly from the southwest: There is nearly always a touch more south or west in it. The hill picks up this slight bias and exaggerates it.

Near the hill, its lee effect and wooded slopes mean the winds are light and variable. Farther downwind the split winds join again and the hill's effect is forgotten. But at my stopping point, I can sit where the two winds choose to conjoin. I habitually say farewell to the southerly and greet its westerly sibling as I begin walking again.

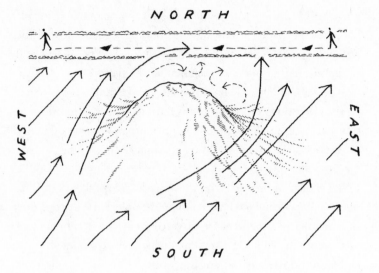

The Split Wind

THE REBEL WIND

Theophrastus described the way local features create winds that whirl. Whenever the wind passes an obstacle that juts into its path, we can think of it being tripped—it turns turbulent. These spinning areas of wind, just downwind of obstacles, are known as eddies, and they're everywhere: The dust spinning by the corner of a building, the chaotic journey of bonfire smoke, and the swirling gusts we feel sitting outside a café as a car passes. And they are acting on every scale, from a few inches to hundreds of miles.

Turbulent wind flow is interesting in so many ways. It offers some certainty. We can say with absolute confidence that once a wind passes a sharp or jutting obstacle, its smooth flow will be broken and a complex set of whirling vortices—eddies—created. Always.

But it also creates uncertainty: Predicting the precise position and behavior of eddies is too complex for science in the real world, and possibly always will be. This relates to a scientific principle known as the "butterfly effect," a sub-branch of chaos theory: The slightest change in any of the variables at the beginning will lead to significant and often total change in the patterns we observe. The proverbial butterfly flaps its wings and changes the weather on the other side of the world. It is hard to visualize the effect on this scale, but smaller examples happen around us every day.

A couple of years ago I enlisted my younger son, Vinnie, to help me study the way the rising smoke from an extinguished candle would change as he walked on the opposite side of the room. The smoke's shapes changed lots and unpredictably. I also learned that experiments need a little more va-va-voom to hold the attention of an eleven-year-old in the age of first-person video games.

The more we learn about the world, the more we discover that the combination of precision and vagueness is helpful because it reflects an inherent aspect of nature. We can describe how many molecules we will find in a quart of air. We can predict exactly how it will behave when heated or compressed. But nobody on Earth could pick one of those molecules and tell you the exact path it will follow for even a second.

It helps if we learn to celebrate this partnership of precision and vagueness when we encounter wind eddies, and this is true on all scales, from mountains to trees. There is not a lot to be gained by knowing that the eddy pattern downwind of Douglas firs is different from that of spruces (but, since you ask, there is a "symmetrical wake" downwind of the firs and a "sinusoidal" one after spruces). We do better to gain a good feel for broad patterns without worrying if there are micro-patterns within the broader forms that will tease, confound, and try to confuse us. We can recognize a person's face and divine the meaning of their smile without having to map every wrinkle in their lips.

The most dependable eddy pattern is a reflection of their whirling nature. The turbulent circular flow of air downwind of an obstacle means there will be places where the wind flows in each direction, including the opposite direction to the main wind. This is what creates the rebel wind, the James Dean of fluid patterns. (It is the rebel wind that creates the rebel dunes we met in chapter 9.)

If anything upwind of you could be described as sharp, jutting, jagged, rough, or angular—anything but gentle and smooth—there will be wind eddies near you. The wind direction will change over short distances and be hard to predict, but there will be places where the wind blows against the broader main wind, the one carrying the lower clouds. If you find this spot, notice how the wind direction and strength you feel change by the step and the second.

Eddies can form vertically or horizontally. Wind that passes over a house roof will roll down into a vertical wind eddy; one that passes a house corner will roll around into a horizontal eddy. In both cases, you'll feel a rebel wind blowing toward the house, even though you're on the downwind side. In large land forms, one type tends to dominate: A vertical eddy will come down over mountain ridges, but a horizontal eddy will roll around any coastline protrusions, such as peninsulas.

The larger the obstacle, the greater the eddy. The more abrupt the shape of the obstacle, the bolder the eddy. If you are downwind of massive, abrupt changes in landscape, like scarp faces or cliffs, and a strong wind is blowing, the eddy will be big and powerful. There may be broad areas where the rebel wind dominates, blowing toward the scarp or cliff, even as the clouds overhead race in the opposite direction. The formal name for the effect on this grand scale is rotor streaming. If you suspect you're in an area that experiences these bold rebel winds, then look for any trees or lower plants. If you're right, they will confirm this trend, pointing toward the wall, even in areas where this goes against the prevailing wind. Rotor streaming on an even larger scale creates the rotor clouds we met earlier (see page 163).

During gales, the effect can be strong enough to fell trees, and if you come across an area of trees that have been smashed to the ground, blown toward a sharp higher crest, it is fair to suspect a rebel wind.

A rebel wind has flattened part of a forest in the Scottish Highlands.

SUMMIT WINDS

Who hasn't had the feeling that the wind wants to pluck them from the top of a hill? We expect it to strengthen as we climb because it is freer with each foot gained in altitude, slowed less by the friction of the surface of the Earth. In 1986 a gust of 173 mph (278 km/h) was recorded at the summit of Cairn Gorm in the Scottish Highlands, but on the same day it reached only 105 mph

(169 km/h) at a station five hundred feet down the mountain and 63 mph (101 km/h) in the valley below that.

Our lowland perspective means we see this as a "summit-strengthening" effect, but actually the summit wind is a truer or more honest main wind: The wind we experience lower down is the one that has been dampened.

Altitude, shape, isolation, and proximity to the coast all influence wind speeds on mountains. The summits with the strongest relative winds are the tallest, narrowest, most isolated peaks nearest the coast.

Where there is a ridge line, there will be significant variation in wind speeds over short distances. If the wind blows along a ridge, parallel to it, there will be fluctuations, but if it blows across the ridge, perpendicular to it, look out for dangerous volatility in its speeds, from half to more than three times those of the main wind nearby.

The wind direction also changes with altitude. The higher we go, the lower the friction, so the less the winds will have backed (fallen left into the plughole; see page 51). Put another way, expect winds to veer as you climb mountains in the northern hemisphere.

THE MOUNTAIN-WAVE WIND

We have seen how lens clouds can form at the crests of mountain-wave winds. Wherever we see them, there will be a huge variation in wind speeds on the ground near them.

The winds on the ground below the clouds will be relatively weak because these are the spots where the wind is riding highest. But halfway between these clouds there will be localized ground winds that are much stronger than nearby because this is where the wind dips down.

The effect is further strengthened when there is a temperature inversion: This acts as a cap, squashing the wind and forcing it to the ground. The Palmdale Mountain Wave knocks down trees and rattles windows in Palmdale, California.

So far, we have considered the mountain-wave wind on a grand scale, which seems fitting, given its name. But we must try to keep in mind that all of these characters come in many sizes. We met a miniature version of the mountain-wave wind in chapter 5, when we saw the saplings that struggled to grow on the wind-scoured bare earth between two woods. A weak rebel wind made an appearance in that chapter, too. No need to go back now, but if you read it again in the future you will recognize these smaller characters. I didn't seek to hide them in the text, in the same way that nature doesn't hide them in a landscape. They're there if we choose to look for them, but not if we don't.

So far in this chapter we have met the characters that are sculpted by the shape of the land. Now it is time to meet a new family, the winds formed by significant temperature differences. The best-known of these is the sea breeze.

THE SEA BREEZE

In November 1950, an elderly patient named Edward Nevin was recuperating at his San Francisco home after a prostate operation. His doctors were very pleased with how his recovery was going. Then the patient took a sudden turn for the worse and was readmitted to the hospital, where he died.

The doctors were mystified and ordered an autopsy, which revealed that a lethal bacterial infection had attacked Mr. Nevin's heart valves and killed him. Further investigations found that another ten patients were ill with the same bacterium, *Serratia marcescens*. The source of the bacteria was a mystery, but the finger of suspicion pointed to the urology department of the Stanford University Hospital, where all eleven patients had undergone operations. The cluster of serious illnesses from an unusual bacterium was written up and deposited in the archives of the American Medical Association. The cases stopped as suddenly as they had started, and the root cause puzzled doctors for two decades.

The truth behind the death of Edward Nevin came to light thanks to declassification, professional investigation, and insight. In 1976 the Nevin family learned of some germ-warfare tests that the US military had conducted in the San Francisco area around the time that Edward had died. The state started to declassify the relevant military documents in 1970, and in 1976 a journalist ran an article that flagged the possible connection between the San Francisco germ-warfare tests and Edward Nevin's death. The medical case files were reopened, and Ed Nevin III, the deceased's grandson and a medical-malpractice attorney, took on the state.

The germ-warfare tests were undisputed. Nobody denied that bacteria had been sprayed from a military ship offshore. The question that needed answering was whether the bacteria had reached Mr. Nevin and the patients in the Stanford University Hospital. It would be answered not by military, legal, or medical experts, but by a weatherman. William H. Haggard was no ordinary weatherman: He was the director of a national meteorological body and a forensic meteorologist. He stepped into court when serious weather mysteries needed solving.

Haggard was able to explain that the germs' journey from the ship to the hospital happened when "an abrupt onset of an afternoon sea breeze pushed the concentrated cloud of particles inland and upward over the hills of the city rapidly as they moved inland."

When the verdict came, the government got off on technicalities that exempted it from legal responsibility. But Haggard remained convinced that the tests had caused the death of Edward Nevin and pointed out that it would have been quite possible to test the germs well away from a populated area.

Sea breezes are not normally defendants in germ-warfare trials. They usually have a happier countenance, bringing cool air to a warm day by the coast.

When it is sunny and there are light breezes—classic high-pressure weather—local winds are stirred into action by differential heating. On a sunny day, the land warms up much more quickly than the sea; the pressure difference leads to air flows from sea to land, and this is the sea breeze. It only kicks in once the land has warmed to at least 9°F (5°C) more than the sea, which in practice means it blows from midmorning to late afternoon. If a warm coastal day starts with few clouds and no breeze, and you feel one blowing in from the sea between midday and late afternoon, a sea breeze is the most likely cause.

The sea breeze is part of a circulation: The cool air flows in from the sea near ground level, which means we can feel it, but then it meets a soft wall of thermals over the land. The thermals force the breeze upward at a point known as the sea-breeze head. The air then flows back out to sea to replace what has flowed inland as the sea breeze.

We can't see the full circulation, but we can spot clues to it in the clouds. There will usually be some cumulus clouds just above the head. In theory, sea breezes can reach more than 30 miles (50 km) inland, but they lose much of their speed as they head over land. You are most likely to feel them within a couple of miles of the coast.

Where I live, roughly 6 miles (10 km) inland, I see the cumulus clouds above the head much more often than I feel the breeze itself. Once you know how to recognize these clouds they're hard to miss, as there are two big giveaways. First, they form a line roughly parallel to the coast and there will be a clear area, with no clouds over the land, between these cumuli and the sea. A sea breeze head is like a miniature cold front.

Second, the height of the head clouds gives a clue as to the strength of the breeze: The taller they are, the stronger the wind. Sea breezes tend to strengthen as the day progresses because of the greater heating. If the sea-breeze front hits rising ground, the effect is compounded: It can trigger more instability and towering cumulus or even set off storms.

The sea breezes that penetrate far inland bend to the right—veer—because of the Coriolis effect. Where sea breezes meet

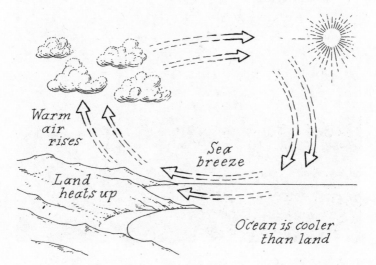

Warm air rises

Land heats up

Sea breeze

Ocean is cooler than land

over a peninsula, the two fronts can join to form a line of clouds, roughly over the center of the land, an aerial backbone of cumulus clouds.

In summer, a fog that appears along the coast and doesn't seem to fit the weather of the day has most likely formed over the sea, then been dragged in over beaches by a sea breeze.

THE LAND BREEZE

The sea breeze has a sibling, the land breeze, powered by the same physics. As we have seen, the heat radiates out of the land at night and especially quickly under clear skies. The colder, now denser air over the land flows out to sea, and the same patterns apply in reverse. A land-breeze front appears with cumulus clouds, but this time out at sea.

The land breeze keeps antisocial hours, partying hardest from midnight until dawn, so you'll meet it less regularly. But

Land breeze after a cold, clear night forms cumulus over the distant sea.

in coastal areas it is common to wake on a clear morning to find a line of cumulus clouds at sea, echoing the coastline.

In the hottest parts of the world, where mountains roll down to coastal cities, land breezes are popular: They clean the urban air overnight. If you spot signs of either a land or sea breeze, take deep, even breaths of the air. Each changes its smell in a way that is intriguing and pleasant. The maritime scent you enjoyed over dinner has been replaced by mountain pines at breakfast.

MOUNTAIN, VALLEY, AND LAKE BREEZES

Sea and land breezes are the neatest examples of local winds driven by local temperature differences. But the same effect is happening to some degree in every landscape all the time. A measurable breeze will flow from a field of well-watered crops toward the neighboring bare earth that heats up much more quickly.

If high ground at the head of a valley faces the sun and heats up more quickly than the shaded lower parts, a flow from cool to warm kicks in, with a breeze heading uphill toward the higher, warmer ground: a valley wind. And if one side of a valley warms but there is cold air below it, again, the breeze will flow uphill from cold to warm. So, valley winds can flow along the valley toward the summit or up one side of the valley that is in sun. (Channel winds are not the same as valley winds. The former are existing winds channeled by the land; the latter are set in motion by heat differences in a landscape.)

These warm, rising breezes can be strong enough to create cumulus clouds above or near the summit of mountains. The dark, steep slopes of the volcanic hills in the Atacama Desert in Chile warm quickly in the morning sun, and cumulus clouds bubble up, the only white in the deep blue of the desert sky.

At night the same forces lead to breezes in the opposite direction, flowing downhill, known as mountain winds. To remember the difference, keep in mind that most winds are named after where they come from, not their destination. Valley winds flow up *from the valley* while mountain winds flow down *from the high ground*.

There is rarely perfect symmetry in these winds. The daily warming winds that flow up the valleys may reach about 12 mph (20 km/h) but will rarely harm anything in their path. The cold mountain breezes that flow down late at night can carry very cold air, slowly killing anything frost-sensitive in their path. If you see frost damage on only the uphill side of plants, they may have been touched by the lethal fingers of a freezing mountain wind.

It is possible for a mountain to experience both a valley wind and a mountain wind at the same time. Shortly after sunrise, a cold mountain breeze may still be flowing down from the summit, even as the sun warms the eastern-facing side of the valley and sets a breeze climbing there.

Large cold lakes and mountains warming in sunshine can combine to create dependable breezes that follow a timetable dictated by the sun's journey. Early in the morning, the Pelèr blows from the north over Lake Garda, then later in the day, once the sun has done its work, the Ora blows from the south. Traders on sailing boats used this breathing rhythm for centuries to commute by the lake breezes. The more famous Lake Como has its own versions, the Breva and the Tivano, but here, any trading boats have long been replaced by vacationers, windsurfers, and posers.

Meteorologists describe rising local winds as anabatic and sinking ones as katabatic, from the Greek for "up," *ana*, and "down," *kata*. The most extreme katabatic winds occur when air over snow grows much colder than the surrounding air and rolls down the hill, picking up speed. Blasts of strong katabatic winds once hit a small yacht I was sailing off the west coast of Iceland. They pushed us over and made for a rough ride.

All of the thermally driven winds above are strongest during warm, sunny, high-pressure weather, when clear skies allow the sun to create the greatest differences in temperatures.

JOINING FORCES

Of course, none of these winds have vowed to act solo. The cold air on a mountainside can roll downhill, join forces with a land breeze, and find itself channeled by a river valley on its way out to sea. Many of our Ancient Greek heroes, including Theophrastus and even the mighty Aristotle, noted very rough patches of sea near the mouths of rivers. Apollonius wrote, "The rolling swell of the sea had been aroused by the strong breezes which blow down from rivers in the evening."

We may know that river valleys are funneling mountain and land breezes on their way to the coast, but I find the effect more memorable when I think of the rivers bellowing at the sea.

WINDS WITH NAMES

There is romance to winds with names, and I adore the playful language and traditions. The ones we met over Lake Garda are far from alone: the Kapalilua in Hawaii, the Virazon of Chile, or the Datoo that sweeps over Gibraltar . . . it's hard not to fall for these characters. Some are so deeply ingrained in the soul of an area that they form part of the culture and lend the inhabitants a sense of place and direction. In France and Italy people talk about "losing the tramontana," just as we might say someone "lost their way" or "went off the rails." Molière used it in his play *Le Bourgeois Gentilhomme*.

Many of the named winds rage and are fearsome enough to harden hearts and minds, more Heathcliff than Mr. Darcy. They change the architecture, too: In many regions, walls are thicker on the side the wind blows from, like those in the path of the Bora, a cold, angry wind that menaces all on its way from the Alps to the Dalmatian coast. Homeowners living on its path lay *golobica*—literally, "little doves"—small, heavy rocks, on roofs to prevent the Bora from stealing the tiles. You can map the route of the Bora by following the stone doves.

I don't want to quash any of the romance, but it's worth keeping in mind that giving a local wind a name doesn't make it unique. Every named wind has its roots in the causes we have looked at. The Fremantle Doctor in Western Australia, for example, is just a sea breeze.

THE CROSSWINDS RULE

As Thomas Hardy wrote in *Far From the Madding Crowd*,

> The night had a sinister aspect. A heated breeze from the south slowly fanned the summits of lofty objects, and in the sky dashes of buoyant clouds were sailing in a course at right angles to that of another stratum, neither of them in the direction of the breeze below. . . .
>
> Thunder was imminent, and, taking some secondary appearances into consideration, it was likely to be followed by one of the lengthened rains which mark the close of the dry weather for the season.

Hardy, like Homer and many other greats since, sets his tales in a natural world that rings true. We're more likely to believe the intricate inventions of love, hatred, and revenge if they are set against a world we recognize, even subliminally. In this case, bad weather follows soon after clouds were seen moving "at right angles" to those at another level. This is probably the best literary demonstration of an important sign known as the crosswinds rule.

The practical part is quite simple. Stand with your back to the low main wind (the wind above tree level, moving the lowest clouds) and note the direction that the highest clouds, probably the cirrus, are traveling. If it is from left to right, a frontal system and bad weather are probably on the way. If the lower wind and upper wind are aligned, no change is imminent. If the upper clouds are moving from right to left, improvement is more likely than deterioration.

How does this work? The crosswinds rule is an odd beast. The theory doesn't fit neatly into any one chapter, as it contains low winds, high winds, fronts, clouds, and pressure systems. But it's mostly about sensing winds, so we'll meet it here. It's easy to use and hard to explain. Fasten your seat belts, as they say.

When a frontal system is approaching, there will always be a mismatch between the upper and lower winds. We know that winds circle counterclockwise around a low-pressure system. With our back to the wind, the low-pressure system is on our left and any high-pressure system is on our right. It is a relationship known as Buys Ballot's law (named for a nineteenth-century Dutch meteorologist). The upper clouds allow us to gauge the wind direction at a much higher altitude.

Here is a weird analogy: You are an ant sitting between tall blades of grass on a lawn. In the distance you can see a gardener with a lawnmower moving toward you. You gauge the low breeze and it's coming from the same direction as the gardener is walking. No problem. Then the lawnmower passes over you and the wind you feel changes suddenly: Now it's coming from a different direction than the gardener. Trouble ahead! In this strange analogy, the gardener is the jet stream and the lawnmower is the low-pressure system; the rotating blades are the fronts. The blades pass over you and you live to tell the tale, but they cause a lot of wind changes and "bad weather."

When used in conjunction with the cloud clues, especially the progression of cirrus then cirrostratus, the crosswinds rule is robust.

The solid theory above that underpins the crosswinds rule can complicate a simple practical method. In summary, with the low main wind on your back, look to the cirrus clouds and remember that if they are moving from left to right it's a clue that a frontal system is on the way. Or: "Left to right, not quite right."

The crosswinds rule works only when you're comparing the main and high winds, not ground winds. It doesn't work with sea, land, mountain, or valley winds.

We leave the winds, briefly, with the thought that every day will bring its character. Don't imagine that the sun can rise and set without one passing you. Many winds try to sneak past us while our attention has been seized by something electronic. Our senses do not make great leaders—they can be childish, faddish, eager for the next bright thing. If our eyes try to lead the day, remind them of who is boss: Feel, smell, and listen to the air. Once we're sensitive to the winds, every day is full of intrigue.

The Trees

H ORSES GROW JUMPY on entering woods. Some
children do, too. The lack of light and sudden confine-
ment sprinkle apprehension and sometimes fear. But so
much of this anxiety turns to wonder when we learn how to explore the microclimates of woods.

In summer you'll notice clouds of insects near the edges of
woodland. They congregate where there is food and water, perhaps something as appetizing as moist dung. But insects are also
sensitive to the way the wind drops near woodland. They like
warmth, which is why you've spotted them in the sunlight and
don't see them a few yards into the woodland. Animals are tuned
in to weather changes in and near woods—for many it is a life-or-
death awareness. We can choose to reclaim this level of awareness,
without looking for dung.

Everyone expects to lose the sun when passing under a tree
canopy, but do we sense how the temperature changes between
our feet and our head? Or how the moisture changes over a short

distance? Do we sense the wind layers? We might notice that woods are cooler than their surroundings in summer and warmer in winter, but do we pay any attention to how much warmer they are, on clear nights, than the open ground? Or that the changes are sudden in natural woodland and gradual in planted groves? Or how the sun rising in the morning can make woods suddenly colder?

Trees influence the weather: They create their own microclimate. But they also reflect the weather they experience. If we understand these two perspectives, the trees will tell us a lot about the weather changes to expect as we enter their world.

THE TREE BLANKET

Let's start with the last question above, as it gives us a good insight into the tree cosmos. Why would woodland suddenly get cooler after sunrise? Imagine you're walking across mixed countryside on a sunny autumn afternoon. You feel the warmth of sunshine on your face and in the air. As you enter the woodland, there are no surprises: You lose the direct sunshine on your body, and the air all around you feels noticeably cooler. If you stop walking for a few minutes you may shiver.

For reasons of science and restlessness, you decide to repeat your route and head out again, this time in the early hours, before dawn. There is a bite in the air under the stars, and a frost is forming on the thistles you pass. Now when you enter the woods, you feel a warming sensation. The woods don't just feel warmer, they *are* warmer. The heat that has escaped from the open country through radiation has been trapped under the leafy canopy. The trees themselves have some residual warmth, and this heat is radiating toward you.

You decide to wait an hour for the sun to warm the land before venturing out from under the trees. The sun rises, and you look forward to feeling the warmth of the first rays. But, to your alarm, there is now a sudden drop in temperature in the woods. What is going on?

The rising sun has warmed the land in places, setting up a gentle circulation of air. The breeze has swapped the warm woodland air with the cold, frosty air outside the wood. It's time to come out from under your tree blanket. The temperature drop has been measured to as much as 7°F (around 4°C). That's a lot if you're already cold, and harsher still if you feel the breeze that carries it.

On a still sunny day or starry night the same effect can be experienced on a smaller scale by walking under a single tree. You will feel the coolness of the shade by day, of course, but it's warmer at night.

There are seasonal fluctuations across tree species, too. Conifers behave in a similar way all year, colder than the surroundings in summer but warmer on clear nights; deciduous woodland is more like open country in the winter and very well blanketed in summer. The rule of thumb with tree blankets is related to the way we thought about frosts. If you can draw an uninterrupted straight line from the ground to the sky, it will be very cold on clear nights.

When setting up a tent at the end of a warm, sunny day, it's hard to imagine that your open area may drop below freezing, even as the forest floor nearby holds much of its warmth. If you're seriously worried about the cold, remember to "wrap yourself in firs." The Douglas fir is much better at holding in the heat near the ground than other species. It might be nearly 10 degrees warmer under a fir than a pine and almost 20 degrees warmer than under a bare oak.

There is little point in looking for dew or frosts in dense woodland: They won't be found under the cover of trees for the reasons I outlined earlier. If there is a good dew or frost on the ground surrounding the woods, it's worth looking at trees in the distance, especially soon after sunrise. A dew or frost may settle on their tops, and when the sun catches it, it will give the woods a fleetingly beautiful sheen. It changes by the second as the moisture evaporates and the sun's angle changes.

The trees also reflect the likely extremes you will experience. Conifers are a sign that an area is prone to severe weather because they tolerate much harsher conditions than broadleaf trees. This is quite a broad indicator, but each tree is also trying to tell us something quite specific.

The holm or holly oak, *Quercus ilex*, doesn't like the cold, but when we see this tree it isn't telling us that the climate is warm. Its message is more subtle than that. Holm oaks will tolerate very cold weather, but not for long periods. They can survive a burst of −4°F (−20°C), but not long periods of even 30°F (−1°C). So if we see a healthy holm oak during a cold snap, it's whispering, "Don't worry, it won't last long."

All over the world, palm trees signal areas where frost can't reach. They're plentiful in the tropics, of course, but in cooler temperate climates they cling to coastlines, where the moderating effect of the sea warms the land in winter. If you can see a few palms in midwinter, however clear the sky is, your forecast is for the temperature to stay above zero.

KNOW YOUR UMBRELLAS

We have all dashed for the cover of a tree when a heavy shower starts. Few consider the different species before they take shelter, but the experience of rain is quite different under each. It's worth giving a little thought to this before the next shower arrives. Indulge yourself in the art of reading tree umbrellas and you'll be itching for it to rain so that you can explore it further.

Tree canopies are not all equal. Spruce, sweet chestnut, juniper, and hawthorn have very dense crowns. Birch, larch, and willow have sparse, open crowns. Scots pine, alder, and oak lie between. But canopy density is only one consideration.

There is a surprising relationship between leaf size and the umbrella effect. The bigger the trees' leaves, the more rain gets through to the ground, which is the opposite of what most people would guess when dashing for cover. During a heavy shower, beech leaves allow twice as much rain through as pine needles. And oak makes one of the worst umbrellas, allowing more rain through than it stops. If you're wondering why or how this is true, the answer lies in leaf architecture. Broadleaf-tree leaves channel water toward their tip—which is why leaves with pointed tips are more common in wet places—but this tip will hold only one large drop before it falls. Coniferous needles are much skinnier, but they can each hold a drop, and there may be a dozen or more needles in the space occupied by a broad leaf. All tree umbrellas work less well in the wind, when the drops are shaken free with each gust.

The lighter and shorter the rain, the more likely it is that any tree's canopy will keep you dry. Eventually they all capitulate and let the rain through, but there are big differences in how each species does this. Some trees are thoughtful to the sheltering walker: They collect the rain at the branches, channel it toward the trunk, and send it down to the ground as small rivulets of water flowing over the bark. Others let it fall away from the center of the tree to the widest parts, in the way a plastic umbrella does. But others let the rain fall straight through the branches onto your head as drips and dribbles. I don't know about you, but I find the constant rain outside the trees preferable to the malicious fat dollops that always find a gap in clothing. Watching rain flow down bark is so much more enjoyable than feeling it flow down your back that it's worth taking a moment to consider which trees are kind and which are mischievous.

The key is to take note of branch patterns. Branches that slope up to the sky or down to the ground make better umbrellas than horizontal ones. The branches on trees that slope downward, like spruces, Douglas firs, and some other conifers, will usher the raindrops away from the center of the tree and your head. During heavy rain, look for a skirt of fat raindrops falling at the perimeter of such trees.

Trees with branches that reach for the sky, like poplars, beeches, cypresses, juniper, some willows, and conifers, will channel the water toward the trunk. Stand next to a beech tree after a good period of heavy rain and you'll witness the torrents picking their way down the smooth bark, leaping out at clumps of moss on the way. (The moss is there because of these rain channels, not because that is the north side of the tree, as many believe.)

Oaks, cedars, larches, and Scots pines have many horizontal branches and mean attitudes. The rain gathers in the canopy until there is too much of it for the leaves to hold. At this point the tree's branches do not lead the excess water inward or outward but let it fall straight down onto us.

The greatest surprise is probably the juniper. The combination of its dense canopy of needles and the shape of its branches make it an exemplary umbrella. It is so good at keeping the rain off the ground that there is sometimes a microdesert under these bushes, with dry-region specialist organisms, like fungi, flourishing there. However, the diminutive size and prickliness of juniper means that few will race to shelter under one.

For combining a dense canopy of needles, a good height and width, excellent branch patterns, and relative abundance, the prize for the best tree umbrella goes to the Norway spruce. During heavy rain, I like to enjoy the dryness of the spruce's umbrella while watching the rain soak the trunks of neighboring trees. It amazes me how long a spruce trunk stays dry during a downpour.

When heavy rain is accompanied by strong wind, rain is blown onto the tree trunks. When walking through my local beechwoods after a storm, I've noticed that the tree trunks wear two different vertical stripes of rain: the one we've just discussed, the channel formed by rain flowing in from the branches; and the painting effect of the wind. Sometimes they overlap, and at other times they follow separate paths; often they intertwine, creating a wet latticework on the bark. It is a pattern that will appear random to all at first, but once your eye picks up the two different causes, the meaning of each is clear and the veil of randomness lifts.

We can enjoy the asymmetric nature of rain in woodland on dry days, too. Notice that the forest floor is never the same for ten feet. There will be a mixture of leaf litter, understory plants, bare earth, stumps, and more. The moisture levels vary enormously for the reasons above, and the dampest spots will often have plant colonies, like mosses, that are very different from those of the drier areas nearby. Compare the forest floor under a spruce and under an oak. The trees that make the best umbrellas aren't good for everyone.

THE WIND BULGE

Wind speeds within the trees are much lower than they are outside or above the canopy, which is what we would expect. Close to the ground, they drop to near zero but never quite reach that until you're touching the forest floor.

As a general rule, the wind picks up with height in the woods: It is much more blowy near the treetops than lower down, but you will also feel a curious anomaly, known as the wind bulge.

The wind strength increases noticeably about one to two yards off the ground and is stronger in this range than a little higher or lower. By chance, this is the height we feel the wind, as our head or hands are normally in this band. There have been many times when I have felt a breeze on my face as I walked beneath branches, then noticed that when reaching up or down with my hand, the feel of the wind fades. It is more distinct in coniferous woods than deciduous ones, but can also be felt in tropical rain-forests.

The wind bulge is probably a result of the lower-branch density between the canopy and the ground, which allows air to pass beneath the leaves. It is the big brother of the tree fan that we met in the chapter 1 (see page 11).

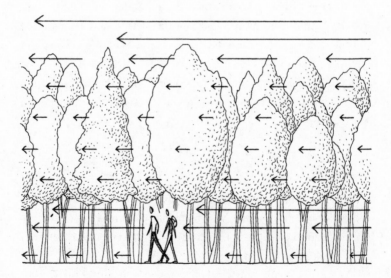

The Wind Bulge

*An aircraft has disrupted a layer of mackerel cirrocumulus, causing a long "distrail."
The cloud has re-formed as a line of icy cirrus below.*

A cirrocumulus mackerel sky

A lenticular cloud reveals a hidden mountain beyond the streets of Inverness.

Visibility deteriorates amid signs of instability in the Sharjah desert. This photo was taken a couple hours before finding rain showers in the mountains.

The author is ecstatic to have found rain in the desert mountains.

The mosses struggle in the dry area under a conifer umbrella.

Mist and rain fall as a warm front passes through. A "wedge effect" in the trees: The prevailing wind is from the left of the picture (southwest).

The rain and wind accelerate past the edge of a wood and flatten the oats.

Bluebells come into bloom first in the warmest patches of my local woods.

Stepped cumulus over the island of Jersey: The wind is blowing from the top right of the picture toward the bottom left.

In summer, the cumulus is thicker and darker over the warm, dark land than over the cool, light sea.

In winter, the warm sea is warmer than the land and is mapped by cumulus.

Birds face into a wind from the left, but this tree's shape reveals that prevailing wind is from the right: a sign that unusual weather can be expected. This was the start of a period of warm, sunny days during a winter high.

Three nearby storms—from the right: growing, mature, and dissipating—are moving from right to left.

There were two rain showers on this beach. A small dog passed before both of them. A larger dog came after both of them. The person walked by after the first, longer shower but before the second, shorter one.

After calm, clear nights in winter, there is a layer of very cold air near the ground. Cars have splashed through the puddle and the water has frozen on the hedge.

WOODLAND HYGROMETERS

You'll recall that the height of the lowest clouds can act as a hygrometer: They descend as humidity increases. In the forest we use slightly different instruments. The first is the simplest. We hear the forest snap, crack, and crunch more when it's dry. The woodland underfoot in February sounds totally different from the same steps taken after a dry July.

The second is the epiphytes, the species that grow on bark, like mosses. The trees are a sign that water is near: There are no trees in the driest parts of deserts. But they also change the moisture levels near the ground and in the air: It is damper in woodlands than outside them. Look at any walls near trees and you will find evidence of damp, the greens and greys of algae and lichens flourishing in the moisture.

Search out an especially damp patch of woodland and a drier area nearby, then look at the base of the tree trunks in each. The more humid it is on the woodland floor, the higher the mosses grow on the trunk. Epiphytes derive all their moisture from the air, so they are a fair reflection of humidity levels.

If you practice looking at this effect, you will have a chance to fine-tune this instrument. Try to notice that there is often more than one species of moss near the base of the tree. We don't need to get out a magnifying glass or look up their names, just recognize that the appearance of the moss changes at a certain height aboveground. The lowest mosses will be the most moisture-sensitive, that is, the thirstiest. The highest moss will be the one best able to tolerate dry conditions.

At the opposite end of the scale, trees can signal drought. Early leaf loss may signal that trees are short of water. If you see a band of trees lose their leaves before others of the same species nearby, they may be struggling to cope on drier soil.

Pine cones open in dry weather and close in humid conditions. They are our third woodland hygrometer. There is a good evolutionary reason for this, and it lies in reproduction. The pine cones are key to the species' survival because they hold and disperse the tree's seeds. The cones' scales flex, closing in high humidity, to better protect the seeds from wet weather, then opening in dry weather, which is better for dispersal.

The pine cones continue to show this response long after they have fallen from the parent tree because the response is mechanical and passive. Humidity causes swelling in the part of the cone scales that leads to their closing, without the need for the tree itself.

THE CASE OF THE MISSING TREES

Trees grow less tall when exposed to wind, which is why they are shorter as we head toward the coast or up hills. Our tallest trees are always well inland and in low country. At the altitude where cold, harsh winds become too much for them to bear, they give up.

In a lowland valley we will see broadleaf deciduous trees, like oaks and beeches. As we climb the hillside, they become gradually shorter and we spot our first patches of the hardier conifers. The combination of short broadleaf trees and tall conifers is a sign that you are at the handover altitude, where the former are struggling and the latter are thriving. We continue higher still. Here, the broadleaf trees have disappeared altogether and even the conifers are less mighty. A little higher, and we find patches of thwarted conifers, nothing higher than a bungalow. We've reached the "kampf zone." Just above this is the "krummholz," where the few wretched conifers that survive show all the signs of extreme weathering, their twisted, gnarled forms as savage as the elements. (The same zone in the tropics has an even more magical name: "elfin wood.") Above this there are no trees, only the lower plants, perhaps grasses with heather and bracken.

We have reached the tree line (or timberline), the altitude at which the exposure is too great for any trees to survive. The timberline altitude varies according to climate, proximity to the coast, exposure, geology, local winds, and many other factors. It might reach as high as 10,000 feet (3,000 m) in the tropics or less than a thousand near the coast or at high latitudes.

The tree line is higher in mountains surrounded by other mountains, and lower on isolated ones. This is the Massenerhebung effect (the Germans really have cornered the market for niche forestry names) and is directly related to the way winds are stronger on isolated peaks.

The height of the tree line is mainly a reflection of climate, but we learn more about the weather by looking at local variations. A natural tree line is never straight. It reflects the prevailing winds and is higher on the lee side of the mountain than the windward. But why does the tree line spike upward at one point or dip at another? The fluctuations tell us about local conditions, especially shelter or exposure. The tree line rises higher on gentle gradients and stops abruptly where the land is steeper or funnels the wind. When the trees reach farther uphill in a valley or struggle on a ridge, they are mapping average wind speeds for us.

All anomalies hold clues, and if you spot a tree line doing something odd, it's always worth a moment's investigation. It's trying to tell you something about the weather in the area. In 1898, the single-handed circumnavigating sailor Joshua Slocum was surprised to find the trees missing from a hillside near his anchorage. In *Sailing Alone Around the World*, he recorded:

> The day to all appearances promised fine weather and light winds, but appearances in Tierra del Fuego do not always

count. While I was wondering why no trees grew on the slope abreast of the anchorage, half minded to lay by the sail-making and land with my gun for some game and to inspect a white boulder on the beach, near the brook, a williwaw came down with such terrific force as to carry the Spray, with two anchors down, like a feather out of the cove and away into deep water. No wonder trees did not grow on the side of that hill! Great Boreas! a tree would need to be all roots to hold on against such a furious wind.

"Williwaw" is a nautical nickname for a violent local katabatic mountain wind. In this case, the cold air above the slopes of Tierra del Fuego was rolling fast downhill to the sea, the same phenomenon that had rocked me off Iceland (see page 203) more than a century later.

WIND DIRECTION FROM TREES

I have looked at the relationship between wind direction and trees in some detail in earlier books because it's integral to natural navigation, so here I will offer a brief recap and add a few new insights where relevant.

Trees are regularly bent over by prevailing winds. The effect is most pronounced in exposed high ground or coastal areas, but it can be spotted in the very top of city trees, too. The rule of thumb is: The less obvious the effect, the higher in the tree you need to look. A hawthorn clinging to the side of coastal rocks will show the effect in its whole body, but a plane tree on a city street might have only a few twigs bent over at the very top.

The tops of exposed trees suffer from "flagging," where the branches on the most exposed side die, leaving only those on the downwind side. This is normally seen only in conifers because the exposure needed is too much for broadleaf trees to survive. The surviving branches, the "flags," point downwind and can be used, like all flags, to give a sense of the wind direction that created them. In this case, the flags don't drop when the wind does: They have a longer memory.

If you see a flagging effect in trees that is confusing, bear in mind that it reflects local as well as prevailing winds. I have been confused on mountainsides in the past when I have found "flags" pointing the "wrong" way, going against the regional prevailing wind. The mystery was solved when I was in the French Alps and realized that the flags all pointed downhill, regardless of exposure. The daytime winds weren't doing any lasting damage, but at night, freezing katabatic mountain winds were ravaging the trees on their journey down from the summit.

Storms blow trees down in two ways. When a tree is wrenched out of the ground in one piece, its rootball twisted out of the soil, it's called "windthrow." This is more common than "windsnap," when a tree's trunk is broken in two by strong winds. They both make truly terrifying sounds.

Windthrow is normal even for healthy trees, but windsnap is usually a sign of some infection and decay in that tree before the storm arrived to finish it off. We can deduce the most common direction that storms blow in from by looking at the trends in both. There is no guarantee for future storms—a gale can blow from any quarter—but the most likely direction is the one that felled most of the trees near you.

Trees are surprisingly sensitive to wind; if they are shaken for only thirty seconds each day they grow to be 20 to 30 percent shorter than a sheltered tree nearby. This is why trees grow short-est on the side of woods that is most exposed to the prevailing wind, which leads to a sloping shape at the edge, called the "wedge effect."

They anchor themselves against the stronger winds by growing roots that are thicker, longer, and stronger on the side those winds come from. The roots can be seen spreading from the base of the tree, and in some species, like poplars, this extends into buttresses that rise up the windward side of the tree, too.

The trees on the windward and lee sides of woods look different. The windward side is hit by a wind that contains lots of particles from the nearby environment, but the lee side is not. If the air is rich with a dust of any kind it will shape which epiphytes grow on the bark of the trees. Maritime air always has an impact, as the salt can inhibit many species, but nitrogen-rich dust carried from a fertilized field may lead to a bloom of lichens on the bark.

As we have seen, the wind will do unusual things near the edges of woodland, where it is whipped into eddies, which are strongest on the windward side. If you see a line of flattened crops near a wood, blame whirling gusts and heavy rain during a gale. It happens on the fields next to our home each year, and the lines are so clear, the damage so total, that it looks as if someone has deliberately trampled the edge of the field. (This has nothing to do with crop circles, which are normally near the middle of fields and must be the work of bored hands or aliens, depending on how you view the matter.) On the flip side, many crops do better downwind of hedges in zones known as shelter belts.

If you study the crops or long grasses that surround an isolated single tree in a field, you may be able to spot the "double-backward-eddy" pattern. As the wind passes on either side of the tree, it is whipped into a pair of "rebel wind" eddies that meet just downwind of the tree and create a wind blowing the "wrong way" toward it. I like to call them the Heart Eddies.

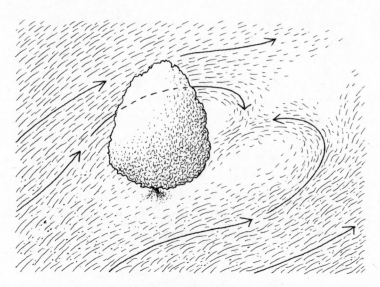

The Heart Eddies

IN THE WOODS

Inside the wood, wind direction is variable, as eddies are never far away. In a typical broadleaf wood, the wind strength at head height—in the "bulge"—is about a tenth of that above the wood in summer and a fifth of it in winter. It will obviously be weaker in dense coniferous plantings and stronger in sparse woods. This is why when you're walking in a wood you can often hear the treetops rustle, even if the breeze you feel is faint. In late summer, if you can't feel a wind at all, pause and look for seeds floating past your nose. They will insist that there is one. Thistle seeds need only the lightest of breezes, about 2 mph (3 km/h), to sail through the sky. Their downy structure makes such a good parachute that it can take them 8 seconds to fall 3 feet (1 m)!

Study any ground and you'll find a map of the very lowest winds, but it is especially rich near trees. Last year's leaves hitch a ride on the wind, and when you see them crossing your path, try to follow them to a resting spot. The land is never flat, and a series of miniature mountain-wave breezes flow over it, sorting the ground into bare patches and collecting troughs. Mosses often mark the bare patches, as they cannot thrive under leaf litter, but near them you will find small hollows where the leaves collect. The effect is especially pronounced in plantings, and I enjoy stepping along the mossy ridges, then letting my feet sink into the leaf trenches between them. If, like me, you enjoy lying on woodland floors every once in a while, you can pick your spot according to the leaf wind map. The bare patches have the breeze; the leaf-filled dips are still.

If you come across a gap—a clearing of any kind—in a wood, the microclimate changes completely: Light levels shoot up, the wind does all sorts of things, more rain falls, and the soil is moister and the air drier. The temperature changes: up if it's day or summer, down if night or winter. The plants and animals change, too.

Campers instinctively pick these spots, but at cooler times of year they are often frost pockets and snow magnets; woodland gaps can receive 50 percent more snow than a nearby open field. Pause to consider whether the gap was caused by a storm or an ax: If a storm, you're in a vulnerable spot if another storm goes through.

SUSURRATIONS AND SENSITIVITY

In *Under the Greenwood Tree*, Thomas Hardy writes of how we can identify trees by listening to their susurrations, the sounds the wind makes as it passes through their leaves:

To dwellers in a wood almost every species of tree has its voice as well as its feature. At the passing of the breeze the fir-trees sob and moan no less distinctly than they rock; the holly whistles as it battles with itself; the ash hisses amid its quiverings; the beech rustles while its flat boughs rise and fall . . .

The science is simple: Every tree has unique leaf and branch features, and the sounds air makes vary as it flows past those signature shapes and sizes. They all sound different. That said, there's camp theater in the field of tree sounds: Another author wrote of an apple tree being a cello, an old oak a bass viol, a young pine a muted violin, and described the "sibilant soufflé of the cedars." I'm not joking. The Japanese listen for *matsukaze*, the song of the wind in pine trees, and they are beyond reproach when it comes to nature and aesthetics. In the fourteenth century, the Chinese scholar Liu Chi managed to combine a musical sensitivity with a more pragmatic perspective: "Among plants and trees, those with large leaves have a muffled sound; those with dry leaves have a sorrowful sound; those with frail leaves have a weak and unmelodic sound. For this reason, nothing is better suited to wind than the pine."

We can achieve this level of sensitivity, if we choose. But how?

It's not difficult to tell a conifer and a broadleaf apart, but to determine a fir from a spruce is challenging. That said, it is possible to capture what birdwatchers call the "jizz" of tree sounds. (Bird experts can perform amazing identification feats, naming a bird after a snatched glimpse of a distant shape or two seconds of its song. After investigation, I discovered there is no trickery, but you have to know the context well.) This is another very helpful and important skill in nature and worth exploring: It will help us recognize the trees by sound.

Imagine you are doing a jigsaw with a thousand pieces. You have barely begun: The four corner pieces and part of one side are down, nothing more. Then a child hands you a piece: It is plain, a smudged white with no discernible detail.

"Ah, thank you," you say. "I think I know where that goes." You place the piece in an open space on the board. The child is mystified as to how you've done it. You point to the picture on the box, one of Turner's landscape paintings, *Newnham-on-Severn from Dean Hill*, and explain. There are dark trees and other land features, some blue sky, grey clouds, white clouds, and a pale horse. All good bread-and-butter Turner. White is either clouds or a horse, but the horse actually has a duskier tint than the clouds. This piece must be part of the cumulus in the distance. It has identified itself by context and exclusion. On its own it looks nothing like a cloud, but that spot is the only option for that color in this jigsaw. In another jigsaw it might be a building or a river reflection, but not this one. That is how expert nature identification works. The sound of the bird or the tree doesn't declare its identity, but it narrows the options to the point at which, if you know the jigsaw well enough, it's easy to place the piece.

A bird expert once explained to me that they often struggle to identify birds they know well when the sound is played indoors. They need to hear it in situ. The same is true of tree sounds and most other sights and sounds in nature. Time spent in any locale is like getting to know the painting on the box. The glimpse or brief sound is the piece you're handed.

What does this mean practically? To begin with, my advice is to get to know well the sounds of the wind in the trees in one patch and treasure them. Don't take on the Herculean task of closing your eyes in lots of different wooded areas and expecting the trees to introduce themselves. It's disheartening and discouraging. If you get to know the sounds of some of the boldest characters in your wood, you'll start to notice how they introduce themselves in new areas, too.

My first experience of this was the clacking of ash trees. Over years, I grew accustomed to the intermittent sound of high ash tree branches clacking together near the edge of woods. Early on, played to me in another environment, I'm not certain I could have placed it confidently, but I knew that it signaled the edge of the woods. To begin with, it was made easier because the sound was always accompanied by the rising volume of the fizzing wind in the trees, the changing light levels, the increase in lichens and ivy, and the change in wildflowers that signaled the same thing. My brain had registered that that particular sound in that situation meant only one thing.

Soon I heard the same sound in the same wood, but in situations I wasn't expecting. The exact sound in the heart of the beechwood without any of the other changes in context triggered a different feeling. Not confident instant identification, but instead a sensation of *That's odd*. I turned around to search the canopy for the cause and an explanation. Ah, there has been a clearing in the past for forestry access and ashes have filled a small gap.

It wasn't long before this sound followed me into new woods. As tree sounds go, the clacking of ashes is distinctive, brazen even. It proved to be a sound that was hard to mistake, even in woods that were new to me, and soon I was picking it out from the more general susurrations of woods I didn't know well. It was the signal in the noise.

This is a very exciting moment, because it is the start of a process of making the impossible possible. Once we have one sound we can pick out easily, we're well on our way, because every sound we recognize fills in the audio picture and brings the others into sharper focus. Next, we add the sound of a gust of wind on beech trees in summer: faraway waves breaking, then rushing up a pebble beach. Then the whispering of poplars. Black poplars are found near water and, to my ear, murmur like water over pebbles. The collection grows until we reach a tipping point where the noise fades and we hear only signals.

We seem to have strayed a long way from the weather, but it has been a deliberately circuitous route and we are now back at an earlier camp. If we practice tuning in to the sounds of each tree in the wind, we cannot help but gain a sensitivity to changes in the behavior of the wind. And, as we saw earlier, any change in the wind means a change in the weather. What starts as a journey of discovery into the susurrations of trees will bring much joy, but also an awareness that makes it impossible for weather changes to creep up on us, even in dense woodland.

Be warned, though: With sensitivity comes a greater sense of vulnerability. Once you have tuned in to the most delicate sounds of a breeze in a chestnut, a surprise gust tugging at the pines above will squeeze adrenaline into the blood and make you shiver on a summer's day. You will stop and turn to the trees, even as your companion walks on, oblivious. You won't believe me until you experience it.

RED MEANS SUNSHINE

As you emerge from the woods, look for red. It's not a color we see very often in the heart of woodland. When we step out from under the trees, the light levels shoot up and many trees salute this with red. The color is caused by chemicals called anthocyanins, which give the rich hue to grapes, blackberries, and plums. In leaves exposed to lots of sunlight, trees use anthocyanins to protect the leaves from damage. I find it helps to think of the chemical and its color as a sunscreen. Young hawthorn leaves are red, as it protects them during this vulnerable stage of life.

And you'll see the same effect in bramble leaves on the sunnier southern side of the bush.

There are many trees that have "copper" cousins, often cultivars, trees specially bred for their colors. Copper beeches look like all other beeches but for a plum, rich rust, or copper color. It is strongest in the leaves that receive the most sunlight and so are more common on the southern side of the tree. Shaded leaves remain green. If you walk around these trees in parks or gardens, you'll notice a good show of colored leaves all around the tree, as some of the sun's light reaches all sides, even if it is indirect. But peer inside the canopy and you'll spot how the rich colors reach deep inside the tree on the south side, but only touch the surface on the north side. All other leaves are green. The copper color is mapping the microclimate within each tree.

Fruits don't color evenly either. For many people, the perfect apple is both red and green. Perhaps this is because those apples balance sweet and sour so well. Or maybe it's because, of the many thousands of varieties, there is a single ancestor, the wild apple, *Malus sieversii*, which has a vibrant mix of red and green. It survives to this day in the mountains of Kazakhstan.

In an experiment dating back over a century, fruit growers placed a cut-out letter on a ripening apple. When they peeled it away, the shaded area behind it was still green, but it was now surrounded by red. It was a simple test that proved the relationship between the coloring of the apple and direct sunlight.

If you spot a red and green apple still on the tree, you will see how the red favors the sunny southern side. I don't know why, but when I see an apple on the ground, it always tickles me to think that I could use these colors to work out which way it was hanging on the tree. And, strange creature that I am, that doesn't stop even when it's in the fruit bowl in the kitchen. . . . I can't help wanting to see if I can put it back the "right way" on the tree. This is the moment I know it's time to head out into the hills again.

CHAPTER 15

Plants, Fungi, and Lichens

Natural Navigation and the Sharper Scalpel • Grass Signs •
Bracken • Annuals and Perennials • Hardiness and Frost Lines •
Time and Temperature • Leaf Clues • Fungi • Lichens •
Blackberries and Possibilities

IN THE OKAVANGO DELTA, BOTSWANA, researchers interviewed communities of small farmers. They spoke to 592 households and asked them whether they agreed with the following statement: "Through certain plants, I can predict whether it will rain or not."

Three quarters agreed.

The Wola in Papua New Guinea study the behavior of a common sword grass they call *gaimb*. When it releases its fluffy seeds, it's a sign that the weather will be fine and sunny for a while. The peasant farmers of Tlaxcala, Mexico, keep an eye on *izote*, the yucca plant, because flowering shoots indicate rain is on the way.

It appears that every culture in the world holds beliefs that plants can help us understand and predict the weather. Closer to home, the scarlet pimpernel wildflower (*Anagallis arvensis*) has many traditional nicknames, including shepherd's weather glass and shepherd's clock, thanks to its habit of closing its petals as rain approaches and late in the afternoon.

These days, few use plants to predict the weather—it is rare now even among country folk—but most hold on to the general belief that it must be possible. We all know that plants react to the weather, so a deep-rooted conviction exists in some that the two are interconnected and that plants can also help foretell weather events. They can. But we must be honest about scales of time and distance. Can a wildflower give us a better picture of the regional forecast over the coming week than a professional meteorologist? No. But it can tell us something about small weather changes we may experience nearby in the coming hours and days that we won't hear in formal forecasts.

The most effective way to use plants to help us understand weather is to appreciate the maps of time and place they make. We start by thinking about climate and season, then narrow our focus to microclimate and micro-season. This reveals, in beautiful detail, the small weather world that the plants inhabit and that we can therefore expect.

Everyone retains an instinctive understanding that climate and plants are closely connected. If we step off a flight in southern Spain, we see fewer grasses, more bare, dry soils, and more pines and palms than when we took off from England. The following year we may land in Florida: plenty of palms but more lush greenery and less bare soil. At some level, even if it is buried, when most people compare the landscapes of southern Spain and Florida, they appreciate that they are witnessing the differences between a warm, dry environment and warm, wet one.

Climate paired with season gives us a likelihood of certain types of weather, but we can never escape probability altogether. The chances of a hot dry day on Dartmoor (about twenty miles north of Plymouth, on the English Channel) in January are low, but not zero. The chances of snow in Florida in January are low, but not zero. It snowed in Miami in January 1977.

Every plant is trying to tell us something about climate and season, and the pair gives us the probability of certain weathers. This is the simplest principle behind the most basic scientific way that plants predict weather. At this level it is common sense, but if we refine that basic approach, some interesting predictions become possible. The past and the present can give us a good idea about the near future.

To start with, we may want to notice only the major shifts in species, how we see more of the thirsty plants on the wetter sides of hills. From a distance, the trees are the first to show this. All pine trees, including Scots pines and black pines, like lots of sunlight, but the Scots pine tolerates wetter ground than the black. Scots pine is more common than black in damp regions, but where the light and soils are suitable for both, and they overlap, the Scots does better on the wetter windward side of hills than the black, which prefers the drier lee side. If the air is humid and you see towering cumulus and other signs of showers approaching, the pines are trying to tell you whether they will hit or miss you.

NATURAL NAVIGATION AND THE SHARPER SCALPEL

Those who already have an interest in natural navigation will be well aware that the plants map the sun, wind, water, and many other variables. Let us warm our synapses by spending a moment enjoying the richness and subtleties of these natural navigation clues: This will hone our skills for when we come to look for weather signs. It sharpens the scalpel of our senses.

We find more sun-loving plants on the south sides of hills, woods, rocks, and anything else that will cast a shadow. In addition, the flowers are oriented toward the sun: In open spots they typically face between south and southeast. At the cold extremes, we find any species on the side facing the equator because this holds on to the last of the friendly climate. That is why grapes in England are on south-facing slopes. But if heat is the problem, we find them on the side facing away from the equator, which is why forests are normally found on the northern slopes of the hot, dry Atlas Mountains (across Morocco, Algeria, and Tunisia). Moving closer, we start to notice that the plants are offering a rich, detailed picture.

For natural navigators there is a fine art within this. Every plant reflects its environment, making both a compass and a map of the area for us. We have to assume that there is a message not just in every plant we encounter but within every part of those plants. Willows indicate wet ground and a possible river nearby, but their catkins flower on their sunny south side first.

Some flowers, like dandelions, survive on both north- and south-facing slopes. They are more common on the south side, and flower on the north later, as the blossom "climbs over the hill." Of course, the blossom doesn't climb anywhere: It just reflects the environment, including the climate and weather it experiences. The warmer, sunnier southern slopes cause the flowers to bloom faster than on the cooler, shadier north side.

If you walked the same route around a hill from mid-March to mid-April, you might conclude that dandelions bloom only on the south side. If you walked the same route in May you might suspect that they bloom only on the north side. Some butterflies, like Edith's checkerspot in North America, will be found on the south side early in the season and later on the north. The dandelions and butterflies are painting the map of the hills, telling us about aspect and microclimate. At this level of sensitivity to habitats and behaviors there is an excellent overlap with weather clues.

We know that rain showers are far from random: They are governed by local heating and the shape of the land. The logical patterns behind them mean they are much more common in some locations than others. All plants are sensitive to moisture levels. If we bring together these jigsaw pieces, we find that certain plants are trying to show us where the rain will fall. Instead of trying to learn all of the species that like wet conditions, take note of the plants around you when you find yourself under a heavy shower. Then take note of where you are relative to high ground, towns, or woodlands. You will start to notice, for example, that certain plants on the windward slope of a hill love regular rain, which you don't see on the lee slope. These plants are trying to tell you where the next rains will fall.

As we saw in the Arabian desert, floral blooms can be triggered in arid regions by rain, and in these areas the flowers map the wet spots for us. This is a subtler, much more frequent example of a phenomenon known as a superbloom, where flowers erupt into bloom soon after rain. In some deserts, plants have evolved to lie dormant in dry conditions for many years, then erupt together after a good wet season in a superbloom. In California this has happened in the past about once a decade in areas like the Carrizo Plain grasslands, about a hundred miles northwest of Los Angeles.

On my walks in damper areas I see a march plant, the soft rush *Juncus effusus*. If it is due to the presence of rivers or large bodies of standing water I can ignore the plant for weather prediction, but if I find it slightly higher up than river level, I question its source. Regular heavy showers are a common answer. If showers are nearby, I know I'm more likely to be hit by them if I'm walking among soft rushes.

How do we know if showers are likely? Alongside all the cloud, wind, and other clues we've been looking at, the plants are trying to tell us. Smells are accentuated when humidity rises, which is why people have reported that the scent of flowers is stronger just before rain. But some plants are vulnerable to rain and they close as it approaches: dandelions, bindweed, buttercups, tulips, crocus, daisies, marigolds, Carline thistles, gentians, red sand spurrey, and wild indigo will all be closed by the time the bad weather hits. Wood sorrel and clover fold their leaves as rain approaches; the wood anemone closes and its stem droops.

The causes and mechanisms vary according to species. Tulips and crocus close when the temperature drops, while dandelions respond to a decrease in both temperature and light. Some flowers respond to changes in humidity, but they are mainly nocturnal, like members of the *Silene* family.

Science and personal experience suggest that flowers exhibit changes before major weather events, but I have yet to come across any evidence that they respond before many of the other signs we have looked at, not least in the clouds and wind. So although plants can help us to predict weather changes, they are not the first of nature's signs.

GRASS SIGNS

Grasses send some simple messages, which is good news because they're plentiful in most habitats. Tall grasses bend with recent and current winds, and sensitivity to these patterns offers one more way for us to stay finely tuned to the all-important wind changes. Grasses that point one way in the morning and another in the afternoon will not stay dry for long.

You're already reading this book, so I need have no fear that going deep will frighten you off. It's best to assume that the grasses are telling us more than we dare believe. They contain secret messages about light, wind, and water trends: They record the weather. Some niche grass clues will really tune the senses.

The common cock's-foot grass moves very slowly toward the wind. Its blades are bent over by the wind, as with all tall grasses, but the new shoots do much better on the windward side of the clump. This is because leaf litter and other debris can gather on the lee side, while the windward is "cleaned" by the wind, making it easier for new shoots to get going. It means the clump of grass creeps very slowly to windward. Its shape will tell you what the wind has been doing today, last week, and last year, and therefore what is most likely tomorrow.

Matgrass (*Nardus stricta*) grows in open, tough environments, like mountain slopes or sand dunes. As clumps grow outward, they flourish on the lighter southern side. Over time, the oldest grasses in the interior wither and die. Together this creates a horseshoe shape in the grass: The two thin ends are aligned to point roughly north and the curved part is at the southern end.

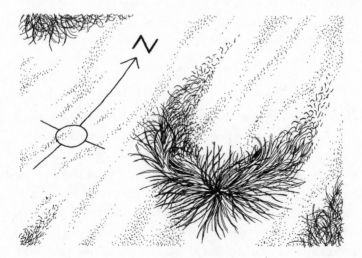

The Matgrass Horseshoe

The lower plants are whispering truths about extremes, too, just like the trees. There are always limits to what each species will tolerate. Frost, flooding, drought, heat, and wind can all kill plants, and plants can die only once, which gives us a different type of forecast. Common heather tolerates cold, rain, and wind well, but is vulnerable to drought: Wherever we see it thriving we can deduce that a dry spell won't last long.

BRACKEN

Bracken, a tall, coarse fern, is a special case. It's a survivor in the broadest senses of place and time. It can be found all over the world and is here for the long haul. In many regions it is considered an unwelcome invader, but anyone who wants to get rid of it must play a very long game. Bracken has been found in fossils dating back fifty-five million years.

Bracken prefers frost-free nights and will mark the frost line on sheltered land. In exposed areas, its height reflects the average wind speeds, getting shorter as speeds increase. In more exposed locations, it gives way to tougher species, like heather and grasses.

ANNUALS AND PERENNIALS

Every region will have species that suggest the likelihood of certain weathers in each season. I won't attempt to list hundreds, but here is a useful rule of thumb: A plant can only tell you something about the season you see it in. An annual, a plant that lives for one growing season only, can survive in places that are pleasant and warm enough in summer but also horribly brutal in winter, because by then the show's over for it. A perennial, a plant that grows for many years, will need conditions to be survivable from year to year.

If a plant is an annual, it is telling you about the spring and summer weather only; it knows little or nothing about the late autumn or winter. Himalayan balsam (*Impatiens glandulifera*) lines riverbanks and damp barren land from the UK to New Zealand and even in some parts of North America. It has large, bobbing, pink-purple flowers and looks like it enjoys warm, gentle summer weather. It does, but it also has a clever strategy for tolerating unbelievably harsh winters. It dies; it's an annual. That is the only strategy that can be recommended in its home valleys— who'd want to be a plant during a Himalayan winter?

Evergreens are perennials, so they make better general winter-extreme gauges. And bulb species hunker underground for much of the year, so they are trying to tell you that things may get bitter. Tulips not only tolerate cold winters but need them.

HARDINESS AND FROST LINES

If you'd like to develop this approach of gauging seasonal extremes from plants, try looking into hardiness zones in your region. Horticulturalists score plants on how tough a winter they can survive. The systems vary across the world, but the principle is the same. The US Department of Agriculture labels plants from Zone 13, for tropical species, down to Zone 0, which means that the plants can tolerate temperatures as low as −65°F (−54°C). For example, tulips are winter hardy in Zones 8 to 3, 15°F to −35°F (−9°C to −37°C).

These indicators are a reflection of climate, but used in combination with other signs, they can help us to predict the weather. In the wild, away from the nurture of horticulture, Nature is harsh. A thousand species might compete for the same square foot of land. This means that every competitive advantage a plant has becomes relevant. We have to assume that each one is telling us something about its niche. Some plants, like dwarf cornel (*Cornus suecica*), thrive only in places with constantly cold nights, so their forecast is simple: Wrap up at sunset.

If we notice plants segue from not-hardy to hardy on a hillside, we have spotted a frost line. With clear skies on an autumn night, we can now see where the extreme cold will kick in. And as we saw with frost earlier, there will be surprises here. If your plan had been to camp in a hollow in October, the very hardy plants you notice down there, like heather, are suggesting you climb above the frost line to where the bracken and other less hardy and more succulent plants are thriving. At the boundary you may see plants that have tried their luck and failed, shrubs that have withered from last year's frost.

TIME AND TEMPERATURE

Last night I spent a couple of hours sitting in a ranger's high seat in a beechwood and felt spring rising through the woods after a sunny April day. The lowest, youngest trees were in full leaf, while the beech canopy had barely gotten started. The smallest trees always come into leaf earlier than the taller ones, a couple of weeks, usually, as it gives them a chance to harvest some light before the canopy trees close out the sky.

Phenology is the study of seasonal phenomena, and we all celebrate its grander sweeps. But seasons don't land on a region all at once. They march over it at their own speed, marked by a collection of changes, some abiotic, like temperature and sunrise, and others biotic, like butterflies appearing or flowers blossoming. These signs are spotted earlier in some parts than others. In the US, spring and summer invade from the south and southwest, then march north and east over the country. Maine might shiver under quiet snow as Florida celebrates balmy temperatures and loud birdsong.

An isophene is a line that joins all the places experiencing the same biological season. It marks the season's advance over a territory. Viewed over a whole country, the line is neat, but the more we zoom in, the more kinks and jags we see in it. As we noted earlier, the season for dandelions is different on two sides of the same hill. And whatever collection of signs you choose to mark a season, you can guarantee it will hit the top of a hill at a different time from the base.

What does this mean for our understanding of weather? We should consider seasons the same as we do climate: There are micro-seasons, just as there are microclimates. And micro-seasons can sometimes be seen within a single plant. Watch hawthorns mark spring. The branches nearest the ground come into flower well before the higher ones. If we can spot differences within one bush, we shouldn't be surprised when we notice that spring is marked a week later in a spot around the corner. The micro-seasons remind us that we should expect to feel the weather change as we walk around a tree.

Temperature changes trigger seasonal growth and govern growth rates, as anyone who mows lawns will know only too well. It varies with each species, but grass stops growing when the temperature goes below 41°F (5°C) or above 90°F (32°C). In a wilder context, the longest grasses map the warm, moist, and sheltered areas for us.

In woodlands, many lower plants, including snowdrops, bluebells, fig buttercup, and muskroot, hold food stores underground over the winter. They burst into life first in the places that warm quickly in early spring. These patches are quite distinct—in my local woods I can watch each of these plants bloom in a rolling map, from places that warm up with the first rays of spring sunshine to those that don't feel warm until May. The last example, muskroot, does a great job of marking the incestuous relationship among latitude, altitude, and climate.

At high latitudes, with their colder climates, it is found in the lower, warmer areas, but in warmer low-latitude regions it is found at cooler, higher altitudes. I only find it halfway up my local hills.

On mountains, we have seen how exposure limits trees to a certain altitude. The same exposure and temperature stresses limit the range of lower plants but also change their appearance. The yield of farmed grasses drops with altitude in a well-understood and predictable way. Heath rush is a tough grass that grows in wet areas on open heaths, moors, and mountains all over the UK. The flower stalks grow shorter and the number of flowers decreases as you climb any mountain. Similar trends are found all over the world.

The plants are reacting to temperature changes all around us, but they are also changing the temperature we feel. On a calm day, we can expect to feel a change in temperature every time we cross into an area with different plants. Each plant responds to the sun's radiation in a unique way, even on a cloudy day, so they are all heating and cooling by varying amounts. We will see relatively big changes if we walk from sand to heather, but much subtler differences, too. Scientists have found variations in air temperature between neighboring areas of grassland if they are used differently. It will be cooler over heavily fertilized agricultural grass than any wild grass nearby. The farmed grass is cool enough to keep grasshopper eggs from hatching, which is one reason grasshoppers thrive only on wild grass.

LEAF CLUES

There is a world to be discovered in leaves.

Leaves are vulnerable to the elements. They always reflect the broad climate and the specific microclimate and can give us clues about both. But we need to take into account any other major factors, like the soil. Leaves grow bigger in moist, fertile soil—think of the big, flappy leaves in jungles compared with the thin, wiry ones near deserts. Once we have done that, some useful trends help us sense the microclimate through leaves.

Leaves get smaller with greater exposure: They shrink as we walk up mountains. One researcher discovered that the leaves of wolfsbane (*Arnica montana*) halved in size when he climbed from 6,000 to 8,000 feet (1,900 to 2,450 m).

Zooming in, we find more local and specific clues. The higher up a plant we look, the closer to vertical the leaves tend to be, but they don't do this evenly, and there are clues in the asymmetry. Leaves on the sunnier south side of plants point more toward the ground, which orients the surface of the leaf toward the sun. Leaves on the northern side of a tree point outward, tilting their faces up toward the sky where most of the light comes from on that side of the plant.

Go even closer, and you may notice that leaves on the southern side are bunched together and those on the northern are more spaced out. The shaded leaves also start farther from the stem than the sunlit leaves. (In botanical terms, the "internode" and "petiole" distances between leaves increase in shade.)

Leaves are also bigger, darker, and thinner on the shady side of trees. The biggest leaves are found in places that are shady but not sheltered from the rain. Try looking low down on the north side of tall bushes or hedges, where there is plenty of shade but no overhanging trees. Such spots get both shade *and* rain, which leads to the largest leaves. These are the sorts of detailed clues that natural navigators use, but they are weather forecasts, too. Dense clusters of small, thick, light-colored leaves are a sign that you are in a place that has received regular sunshine and will do so again before long.

Closer still, feel the leaves: Their texture reflects the most likely type of weather in an area. If they're especially thick and

internode

petiole

Shade **Sun**

fleshy, then hot, dry weather is most common. If they're a little tough, like leather, cold, dry weather is more likely. The air gets colder and drier with altitude, so the leaves also become more leathery as we climb.

If you walk across a landscape with a wide variety of microclimates, from dry and sunny through dense shade and wet areas, keep an eye on how the plant species change, but also on how the leaf shapes alter. Even the curvature of leaves offers clues to the microclimate. Most leaves that arc, that start pointing up to the sky and finish pointing down at the ground, are found in deeply shaded or wet areas. The effect can be seen in foxgloves and some orchids, as well as many grasses.

FUNGI

Mushroom hunters are finely tuned to the elements because weather fluctuations strongly influence fungi fruiting. The fungi tell us much about the soil, the plants nearby, and the microclimate.

One afternoon my son pointed out an obvious clue that I had overlooked. He was back from school and we were all sitting around the kitchen table when I started describing some puffballs I'd seen earlier. He seemed genuinely curious, not something I expect when it comes to fungi conversations at the end of a school day. I explained that puffballs emit a cloud of millions of spores when nudged by a falling raindrop. I told him we could go into the woods and simulate the rain by prodding the fungus with a thin stick and watch the spore cloud leap into the air.

"So," he said, between mouthfuls of toast, "you only get these fungi in wet places." There was no rising intonation at the end of his sentence. He wasn't asking me, he was telling me. I couldn't have been prouder. I was also a little concerned that my brainwashing was displacing trigonometry or some other knowledge he might need for his upcoming exams.

Fungi are sensitive to changes in light, temperature, and humidity. Many homes conduct experiments to prove this: If you have ever noticed dampness in a building, it is probably not the moisture you spotted but the fungal growth thriving on the cool, shaded, moist surface.

Fungi do not use light in the same way green plants do—there is no photosynthesis—but light is still important, as it signals to a fungus that it has reached the surface, the right level for fruiting. Temperature and humidity fluctuations are more critical. The fruiting time for many fungi is hastened by either a cold shock or a large fluctuation in temperature in late summer or autumn.

There is a widespread belief that mushrooms sprouting overnight indicate increased humidity and therefore bad weather on the way, but the picture is slightly different for each species. It is true that long periods of dry weather inhibit fungi, so a sudden change from very dry to humid conditions will encourage fungi to make an appearance.

It is unlikely that sprouting fungi are the first sign of weather change you notice. Many of the other cues we've met will beat them. Fungi join a chorus of signs: They don't lead the choir. I take more time to enjoy looking for fungi after the first frosts and when I have felt a roller coaster of temperatures in early autumn. But in terms of weather signs, I tie the arrival of fungi to the major changes we feel when colder air masses arrive in autumn. It is more a pairing than a predictor, but still satisfying to notice.

As with so many biological cues, we're only beginning to appreciate the beautiful possibilities in this area. Some scientists believe that many fungi species eject spores in response to a drop in air pressure, and this could, if proved, lead to some extraordinary weather predictions. It hints that fungi are tuned in to every change in their environment and that many exciting discoveries are still to be made in the relationship between fungi and weather signs.

LICHENS

Lichens offer some wonderful weather clues, but before we look for these we should get to know some of the habits of these delightful organisms—especially their relationship with water. Lichens harvest water directly from the air, which makes them purer gauges of microclimate than either plants or fungi. Many lichens thrive by the sea and others won't be found within a few miles of the coast, regardless of soil types in the region, because of their sensitivity to salt and the character of the air.

Lichens, like mosses, are telling us about the humidity of a region, and they do this in two main ways. Many lichens thrive only in damp areas, so if we see these examples growing bountifully, we can be confident we are in a moist environment. But it is always worth looking for the "lichen flag." Depending on the character of the local winds, especially when the prevailing wind is very moist or dry, the lichens will do better on one side of trees than another. Light plays a bigger part with shallow, surface lichens, but with lichens that hang or are bulbous, it is always worth looking for a wind sign in their tendency to thrive on the windward or downwind side of trees or other barriers.

In places where the wind isn't the dominant factor, the fog might well be, and the lichens make a fog map for us. (If you're unsure whether the wind is dominant, look at the very tops of the tallest trees: If they're not bent over, the local winds are modest.) Many lichens love fog: It gives them the precious moisture they need and can't get from soil, without their having to rely on rain, which may be rarer.

Researchers in the Netherlands made an amazing discovery. They figured out that you can tell how many foggy days there will be in a month, even whether they are day or night fogs, by looking at the lichens. (Lichens prefer day fogs because they allow them to shut down at night, which helps them cope better with the driest times.) The next time you're walking up a hill through a morning fog, try to note the lichens on the trees and whether the types change as you emerge through the fog to the drier air. Make a mental note of the lichens that thrived in the fog. Then, on a day without fog, you can use them to predict and map the areas that will be foggy next time, too.

Many lichen species react dynamically to moisture levels in the air, swelling as it grows damper, as the naturalist Richard Jefferies noted in the nineteenth century: "Black-spotted sycamore leaves are down, but the moss grows thick and deeply green; and the trumpets of the lichen seem to be larger, now they are moist, than when they were dry under the summer heat."

BLACKBERRIES AND POSSIBILITIES

Many natural wind maps are well hidden beneath our noses. Late summer or early autumn are wonderful times to forage for wild fruit. Track down some blackberry bushes with plenty of ripening fruit and begin tasting. You will soon notice that not all the fruits taste the same: There are some quite striking differences.

There are over two hundred subspecies within the blackberry family, and this accounts for many differences from one place to another, but within the same area one of the starkest contrasts is between sunny and shady areas. Sweetness comes from the sugars in the fruit, and this energy comes from only one place: the sun. We get sweeter fruit in south-facing spots open to the sky.

But this is probably just plucking the surface of what is possible. Research in South America with a different species of fruit from the same family, the *Rubus glaucus*, is turning up some extraordinary results. It is not the same fruit as our blackberry, more akin to a loganberry, although it looks vaguely similar. I mention it here only because it offers a glimpse of what might be possible if we take the time to get to know our local plants better.

The Latin American blackberries make a banquet of tangy fruit *and* a wind-speed gauge. Each berry is actually a collection of much smaller "fruitlets"—each of the tiny bulges we see on the berries can be thought of as an individual fruit. Counting the number of fruitlets in each one will give you a measure of likely wind speeds where you're standing. The more fruitlets, the higher the average wind speed.

How does this fruity anemometer work? The blackberry is pollinated by insects, but it is also self-pollinating by wind. The windier it is, the more pollination is possible, and so the more fruitlets per blackberry.

Plants always reward habits of curiosity and observation, building layers of meaning. The signs are always there because nothing is random, but there is no pretending that they're all obvious. We may see some palm trees in Ireland in summer, get the message about mild winters, and feel we're in a warm microclimate. But winter mildness is not the same as summer warmth or dryness. What are the chances that we also noticed fields of barley all around us but none of wheat? This is a sign that rain is never far away.

Southwestern Ireland is mild enough for palms to grow, so we might assume that wheat would thrive there, but it doesn't: It's too cool and wet in summer. In truth, southwest Ireland is so wet that there is a different way of using the crops to forecast the weather. If you can't see the barley, it's raining. If you can, rain is on the way.

If you pass a field of wheat or barley, you might like to know that in Japan, *honami* is the study of how crops form waves in the wind.

CHAPTER 16

The Hoodoos: An Interlude

T HE DAY HAD STARTED in a worrying fashion. At a minute past eight that morning I had been sitting in a cosy café called Tooloulou's in Banff, Canada. Old wooden chairs surrounded high tables that were covered with laminated gingham. The air was thick with syrup and fat. My left hand was tucked into my jacket pocket, nursing a can of bear spray.

I was, I learned that morning, afraid of bears. It didn't matter how often I ran through the facts in my mind, my emotions had little regard for them. I had taken the trouble to establish that there had been no fatal bear attacks in the Banff region in November for decades, but when a horror gets hold of us it will take its claws to such statistics.

Mine was an irrational fear, but so are most of our outdoor anxieties. It's sensible to be cautious in poor light, but we fear things in the dark that escape any calm analysis.

Most accounts of bear-attack survival mention that the victim kept their spray within easy reach. Before getting fully dressed that morning I had practiced pulling mine from my pocket and slipping off the safety catch in one action. I had also reread the National Park guidance for surviving a bear attack. Apparently, when confronted by a bear, we must pick one of two strategies.

If you think the bear is on the defensive, behave in a way that will reassure it. Don't run away, as this may prompt a chase, which, you guessed correctly, you will lose. It's best to edge backward slowly, making soft, reassuring noises—my warmest congratulations to anyone who manages that in the circumstances. If the bear continues to approach and you find yourself cornered, lie on the ground, play dead, and protect the back of your neck with your hands. Please just imagine doing that.

If, however, you think this is a predatory attack and the bear has targeted you as a source of food, then the approach is very different: Fight back with everything you've got, including rocks, knives, and anything else at hand. Who, I wondered, would have either the experience of these situations or the presence of mind to assess an approaching bear's state of mind and therefore the correct strategy? Not I.

A young woman in the information center explained that my spray was effective up to a range of about 15 feet (5 m). That sounded a lot to me, until I thought about it. I pondered whether I would have the courage to wait until a bear drew as close as a car length away before pulling the trigger. Would I really dare wait until I saw the whites of its eyes? Do bears have white in their eyes? No, they don't. But just considering that at the wrong moment might lead to a mauling. And if I pulled the spray trigger too soon, would I be converting a bear deterrent into a bear irritant or, worse still, a bear enrager?

My right hand held the menu, which was also part of my bear-defense strategy. I've heard it said many times that hungry bears and knapsack food are a dangerous mix. My solution was to walk for the day without any food on me. Heading into snowy hills without any food is poor practice but, not for the first time, I was tailoring outdoor wisdom to suit myself. And this cloth was being cut to fit a simple fact: Starvation would take weeks to kill me; it would take a hungry, angry bear mere seconds.

This tailored and questionable strategy demanded an uncompromising Canadian breakfast. I had been looking forward to that North American peculiarity of a broad plate of food that segued from savory to sweet and back again. I placed my order with a waitress, who assured me through an unwavering smile that I had chosen extraordinarily well. In countries with a strong tipping culture, we get to feel wise after the smallest of decisions.

Very soon, a bacon and spinach omelet arrived, nestled next to a heap of blueberry pancakes. Blueberry pancakes turned out to be a euphemism: There were pancakes and blueberries to be found, but they lay under a soft mountain of whipped cream, confectioner's sugar, and maple syrup. I've heard it said that obesity is a multifaceted problem. Some of these facets looked up at me from my plate.

I had been working in Calgary and looking forward to this day off. The plan was simple. I would walk my digesting breakfast to the top of a modest mountain, which at 5,500 feet (1,675 m) would be considered a giant in the UK and a bump in the Rockies. I would be greatly helped by the fact that my start point, Banff, was already at 4,500 feet (1,370 m). After descending, I would do some modest cross-country natural navigation, then follow a different trail to the hoodoos. What are *they*?

Hoodoos are tall, thin rock formations created when a small section of soft rocks is protected from erosion by wearing a hat of harder, heavier rock. They have many wonderful names, such as earth pyramids, tent rocks, and—my favorite—*cheminées de fée*: fairy chimneys.

My sensitivity to the wind during the walk would be as high as it had ever been. I studied the light, the clouds, the trees, and the shape and color of the land. I felt a mountain breeze rolling down from above, cooling my face at the start of the day. The conditions were right for looping: My scent would be climbing and descending.

Crossing an opening between two forests, I relaxed as I felt the gap breeze on my back: It carried my scent ahead of me, and it is difficult to surprise animals with a working nose if you walk downwind. And each eddy was a source of pleasure, too, spreading my scent in a wide circle—the smell equivalent of "Coo-ee! I'm here!" But inevitably there were times when I had to walk with the breeze on my face for long periods, and I didn't enjoy it. The conditions were right for me to stumble on animals, a joy at most times but not today.

I followed a trail uphill, watching the conifers grow shorter. At the summit, there were fewer eddies and the winds were stronger—I saw their power etched into the trees all around me. I looked out over the Rockies and spent a happy half hour letting my eyes roam along the distant tree and snow lines. The tree line was noticeably higher on the sheltered side of the mountains that I could see. In three steep parallel gullies the conifers charged a few hundred feet higher than their siblings on the exposed ground at either side. There were a hundred snow pockets in hollows below the main snow line.

On the walk down, I listened to the wind in the pines. The notes fluctuated as I rounded the rocks, and the volume dropped the lower I descended. Then something about the sound made me stop near a bend in a trail. Or did I imagine that? There, between two lines of pines, was a pair of electrical lines stretching away to a pole in the middle distance. Had I picked up a change in the sound because of those wires? I don't know.

After a few hours on trails and off, the pancakes were a memory and I had seen no signs of bears. If facts do little to dispel fears, physical exercise and a little hunger work wonders. Suddenly a snack in the rucksack seemed a risk that would have been worth taking. After all, I was only a few miles from a busy town. Then I heard it. A bark? A groan? Something animal, loud and not far away. I froze. And then I began to sing.

Singing is one of the bear-deterrent strategies encouraged in the area—honestly! It's supposed to be a uniquely human signature that the bears identify easily and move away from. In other parts of the world, like Scandinavia, experts advocate wearing bells on knapsacks, the constant tinkling advertising the approach of a hiker to the bears, which then have plenty of time to leave the area. But I had been persuaded by one old hand that in the Rockies this was not the way to go: Apparently, bells sound like the tinkling of water to grizzlies and they will approach to investigate.

Investigative grizzlies I could do without. Being tone-deaf and with a range of only a couple of notes, there are few songs that I can sing confidently, and for this reason I know one of them by heart. "Sixteen Tons" by Merle Travis rolled unmelodiously into the trees.

Launching into the fourth verse, I was serenading the pines with lyrics about one fist being made of iron, the other of steel, when light levels leaped up and my singing weakened. I stepped out gingerly as the trees gave way to a meadow. I felt the temperature drop.

I struggled to focus on the scene. Wood blindness after only a few hours in the trees would be to exaggerate the problem, but focusing was a challenge for a few seconds. With my hand back on the bear spray, I watched brown shapes move in front of me. It was hard to tell if they were large and distant or modest and closer. Somewhere in between: I was looking at a herd of elk. About a dozen were grazing on the lush grass about seventy yards from me. They were impressive, like great fallow deer on steroids, every part grander and more substantial than the ungulates I see near home.

I stood watching them for ten minutes. They looked up at me from time to time but were unmoved by my presence. The day before, I had heard stories of how elk had charged people recently in this region, probably worried for their young, but I didn't feel the slightest concern. Their overblown similarity to other grazers gave me a sense that I could read their body language well enough, and they, hopefully, mine. If I stuck to the trail at the edge of the grass, I felt sure we could enjoy each other's company without raised pulse rates on either side. The cold crept inside my jacket, and I reluctantly left the elk to their meadow.

The Hoodoos

The trail led all the way to the hoodoos. To reach them I had to scramble down soft rock, mud, and scree. It is one of the immutable laws of walking on gradients that a section of steep downhill is an entirely different experience for our feet from steep uphill, even though we are often walking over exactly the same ground in each case. Every few seconds I felt a foot slipping and lowered my center of gravity, grabbing extruding roots where I could and watching soil race down ahead of me.

The hoodoos looked vast from a distance and less formidable close up. When I stood at their base, the honey-colored stone rose only a few yards above me. They reminded me of termite mounds, albeit taller and skinnier. I sat on the pale earth and spent a happy quarter of an hour staring up at the sandy monoliths. I sketched them, photographed them, and then scrambled back uphill to contemplate their forms from level ground.

My route back to town took me past a parking lot and a knot of tourists who looked like they'd seen a Canadian breakfast or two. After the reverential serenity of my morning wanderings, I found their loud conversation hard to cope with and took the longer route home, heading away from the road and down through the forest toward the river again.

In the woods, I felt a breeze at face level, but nothing a few feet above or below that. I could hear the tops of trees wrestling with something stronger. The mosses climbed higher up the tree bases as I descended toward the water. The moment I dropped out below the trees I felt a channel wind, and it tracked me along the river. Near the water the trees gave up, and there was settled snow on the ground: On the northern side of one great boulder it pointed perfectly north in a broad, tapering arrow. Clouds had covered the sky, but before that the southern midday sun had thawed all the snow on the dark mud, except on the shaded northern side of the rock.

The peak of Mount Rundle soared to a point and towered over the valley. Its imposing dark face was lined and striped with snow.

Over millions of years, sedimentary rock had been bent and contorted into this fine mountain that pawed at the sky. The snow had acted like the dust of a fingerprinting detective, and bright white lines shone against dark stone. Every line, buckle, and twist could be seen as clearly as writing. Then I felt it. This is not *like* a language, it *is* a language. But there is a difference between recognizing something as a language and the ability to understand its meaning. The rocks were speaking, I was listening, and neither of us understood the other.

At the river's edge the sky opened up and the mountains were clear against its mottled blue, white, and grey. The philosopher Friedrich Nietzsche felt vulnerable to the electricity of clouds and the all-seeing sky, and saw forest-clad mountains as a shield against these malign influences on his mind. In Nietzsche's case, it was perhaps a symptom of incipient insanity, but on emerging from the coniferous cover of alpine woodlands, it is a sensation that can be hard to resist for everyday thinkers, too.

The sun came out and warmed my face, and then I felt it lifting off the dark soil by the river. The clouds piled up, and towering cumulus formed. The bottoms of the clouds were rough and ragged, lower than at the start of the day. The tops were well defined but climbing, morphing, and starting to bend with the wind. A picture of likely showers, maybe snow, but no sign of actual storms yet. Stop-and-start weather.

I looked to the mountain peak, hopeful of spotting one of the summit clouds, a lens or banner perhaps, but also aware that the conditions weren't right for either. There were patterns in the way the clouds were being carried around and over the highest ground, a few more clouds on the windward side and more blue sky beyond, but nothing that would warrant its own name.

At my neck's insistence, I lowered my gaze from the sky, then followed the dry riverbed toward town. The snow began to fall again. It was mild, and the flakes were big. Within minutes I was being snowed on by a blue sky. It was here that I saw the most beautiful pattern I had seen in weeks.

Something about the smaller stones by my feet grabbed my attention. I squatted and looked at the pebbles, then dropped to my hands and knees and studied the pattern. Crawling across twenty yards of the rocky riverbed, I winced as one sharp edge cut into my knee, and I rubbed a snowflake from my eyelash, the better to study the stones. I had never seen anything quite like it.

Like so much in nature, the forces at work do not surprise us when we are told of them, yet they mesmerize us when we stumble on them. The river was dry because we were at the end of the dry season; in six months it would be full of fast-moving water as snow melt ran down from high ground all around. As the water surged through this part of the valley, it had acted as sifter, sorter, and sculptor.

Thousands of rounded rocks covered every square foot of the riverbed, and they had been perfectly sorted by size. On the inside of the bend the pebbles were small, rounded, and nearly uniform—each would have fit comfortably in my gloved hand. On the outside of the bend there was a jumble of larger, more uneven rocks, much bigger than my hand. And between the extremes, the flowing water had created a perfect grading system than ran across the breadth of the river.

A gust carried the freshly fallen snow along the rocky riverbed. It settled in the dips between the bigger rocks on the outside bend but raced on quickly and without rest over the smaller, smoother stones on the inside. I watched flakes drop out of the air in the wind shadows of the larger stones. And then, by a large rock at the water's edge, a micro-blizzard took a U-turn and blew the wrong way, hitching a ride on a rebel wind. It was so beautiful it made me quite weak and euphoric. I sat on the rocks for a few minutes, gazing in admiration at nature's art. At that moment I felt connected. I also felt I could have reasoned an aggressive bear into companionship.

My route back into town took me past the Fairmont Banff Springs Hotel, a magnificent beast. There aren't many hotels with aesthetics that inspire, but this one is sublime. I think it's the finest-looking hotel I've ever seen. It could be the prettier, friendlier sibling of Colditz Castle, the infamous prisoner of war camp in Germany during World War II. I pitied the poor souls staying in it because they couldn't see it.

A wooden walkway near the hotel was mostly clear of snow. But there was a little in places and it was not sprinkled randomly. Straight lines are rare in nature, and any that appear pull our eyes toward them. On top of the wooden planks, I saw not just straight lines but intricate patterns: a latticework of straight white lines, painted by a thin layer of snow. It was an artwork that had my full attention instantly. For a moment it puzzled me. Then I felt the sun on my right cheek and it all made sense.

A thin coating of snow had fallen on the walkway overnight. It had settled on the raised platform, lifted off the earth, but not on the stones or soil a few yards away. The air temperature was below zero and none of the settled snow had thawed during the morning. By the time I reached the area, it was the early afternoon, and a warming sun was out. It had sublimated the snow on the boards, turning it from ice crystals to vapor, skipping the water stage. And it had happened quickly, in a few minutes, during which time the shadows cast by the wooden handrails left the snow patterns on the planks. A line from the snow through the handrail made a perfect compass, which pleased the navigator in me. But the snow pattern was better than that: It was unique. A painting that marked a momentary meeting of sun, snow, and shade. I will never see the same pattern again.

What a day full of enjoyable lessons, as each one is, if we show up. I had foolishly allowed the bears to fill my mind as I set out in search of the rock pillars. The hoodoos were magnificent and well worth the walk. But however impressive a physical landscape, the weather is always there, adding a layer to every outdoor experience. If we look for the signs, the weather will reveal its own collection of delicate artworks in the sky and on the ground.

Snow shadow patterns on the Banff walkway

The City

AFTER SIXTEEN YEARS of rural life with occasional trips to London, Sophie and I felt our elder son could do with a burst of intense urban immersion. In October 2019 we flew to New York for a five-night stay in Midtown Manhattan.

Ben had never been to the US before, and Sophie had never been to New York. I had been a few times and, on day one, adopted the tone of tour guide. Sophie knew a good tactic to shake this off, and after a few cultural sorties, she dragged a delighted teenage boy and huffing dad into every shop that had a pair of running shoes or a bright logo in the window. There are plenty of them.

I can walk from to dawn to dusk without complaint in the hills, but two hours of that was too much for my endurance, and I begged early release for good behavior. And so it was that we agreed to split for an hour or two each day and meet up later.

We had a brilliant holiday, and it was made possible by each of us remaining sensitive to our different interests and my very limited shopping talents. The highlight of our stay was racing remote-control sailboats on the pond in Central Park. Sophie found the coffee in the café there excellent, Ben loved the racing, and I enjoyed edging ahead by using the wind eddy on the far side of the pond.

A pillar of steam rising from the sidewalk split us once more. More subtle than the yellow cabs, steam rising in the street is an image that seeps into foreigners' picture of New York because film and TV directors love it. I had no idea before my first visit to Manhattan that steam is a utility in New York, like electricity or gas, and has a long history as a source of heat.

Like any good tourist, I stopped to take some photos of it. But after a minute of snapping, I didn't want to move on. Sophie had spotted something farther up the block she wanted to investigate, but I was transfixed, stuck staring at the steam snake as it slithered upward. Never had a city creature been so keen to show me what the air currents were doing at street level. There was no chance I could let it escape.

MEETING MARILYN IN MANHATTAN

When a wind hits the perfectly flat, smooth wall of a high-rise building it is deflected in many directions, but much of it travels vertically downward. When this wind then hits another perfectly flat surface—the sidewalk and road—it bounces out and upward, creating a puff of wind that feels as though it is rising from the ground itself. This rising air can lift skirts. Hence it has earned the nickname of the Monroe effect, after the iconic scene in the 1955 movie *The Seven Year Itch* when Marilyn's white dress is lifted above her waist when she stands on a subway grate. In cities, there are winds that complete a circle as they hit a building and then head down, out, and up.

I tried to engineer some reason that meant we needed to go to Forty-Second Street, and it wasn't hard. There are a hundred good reasons, and they don't hide their light under a bushel. My secret reason for wanting to walk down this street wasn't listed in any of the guides or featured in any ads. It was not lit in neon or shown in movies. I was on the hunt for the urban-canyon effect.

The gap- and channel-wind effects can work together in urban environments. This is one reason that on windy days we so often feel a strange alternation between calms and repeated sudden blasts of wind when walking along a city street. Anytime the wind is blowing roughly perpendicular to our route, it will accelerate down the streets—they are as good as valleys to the wind—and their direction will be channeled by those we cross. The combined effect is like an on-off switch each time we step from the sidewalk to the road and back to the sidewalk.

An extreme version is known as the urban-canyon wind. If a street has very tall buildings and is unusually straight, the channel effect is exaggerated. We don't get many street canyons in Europe, as the cities have grown too organically, their layouts all twists and turns. Younger cities allowed planners and architects to draw long straight lines. It's a wonder that they were able to draw anything straight, considering how drunk they must have been on the power.

If any street narrows or widens, it has a big effect on the wind strength, because of the Venturi effect. In its simplest form, a narrowing street is like a thumb over a hose—the wind accelerates. In the most extreme version, a wind is squeezed between two buildings and can reach three times the speed it was in the wider part of the street. The result is a type of gap wind.

When the wind blows across the urban canyon, all sorts of eddies form, and the wind can be seen blowing in half a dozen different directions on the same street. The clouds overhead move one way, flags are tugged another, and you feel three different

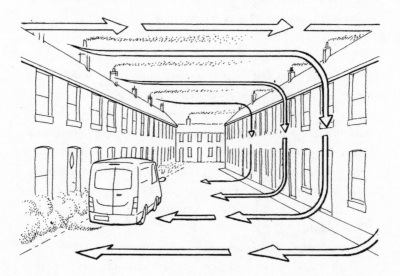

The Pollution Effect

winds as you cross the street. One of the strangest side effects of these eddies is a massive difference in air pollution. The exhausts from cars are blown toward the leeward side of the street, while the downwind side gets fresher air from above.

THE HEAT ISLAND AND THE CITY BREEZE

The other day I was on the phone, speaking to someone in an office in Chichester, the nearest town to our home, who said, "I can see a couple of buzzards from my window."

"Are they circling?"

"Yes."

"Are they looking for a car-parking space?"

"What?"

"Are they over the car park?"

"Yes. How did you know?"

"It's midmorning, and the only place near you warm enough to cause a thermal at this time of day is the tarmac of Northgate car park."

Cities are subject to the large-scale weather phenomena that we have met, but some effects are the city's alone. The millions of tons of asphalt and hard rectilinear architecture form a unique relationship with the elements.

The sun warms the buildings and roads more efficiently than most natural substances, so the city warms more quickly and reaches higher temperatures on sunny days. The dark materials absorb more solar radiation, but the buildings also present a larger surface area. At the beginning of the day, the low sun can heat a skyscraper much more effectively than it can a tree. At the end of the day, as the heat radiates quickly out of the surrounding areas, the buildings and roads act as heat batteries, which now release the heat back into the city. Water that coats trees in the countryside, cooling them through evaporation, runs quickly off steel and glass and down drains in cities. Then there is human activity, the heating, industry, and vehicles. Together, these effects combine to create a "heat island" effect that can make cities 22°F (12°C) warmer than the surrounding countryside.

As we saw earlier, localized heating creates cumulus clouds. They bubble up above cities, then bend or march slowly with the breeze. The heat island also creates its own winds. As the warmer city air expands and rises, the cycle we saw with sea breezes (see page 196) kicks in. Cooler air from the suburbs flows into the city. On a warm, clear, sunny day, with little apparent wind in the early morning, the city starts to make its own, and a gentle breeze will flow toward the city center by midafternoon: the city breeze.

The city breeze will be felt only on days when there is little or no main wind. If a light main wind is blowing, the downwind side of a city will be several degrees warmer than the windward side. But a strong wind removes the heat island effect altogether, which means temperature drops are exaggerated in urban areas. If, after a summer high and a few days of very warm, calm weather, a front goes through, it robs a city not just of the sun's warmth but the whole heat island effect, too. In one day, we go from seeing people's skin in the parks to coats on hunched shoulders.

THE CITY SPLITTER

Cities are divisive: They split the weather that tries to pass over them. From the vantage point of a tall building, you can watch rain approaching your city, only to see it break into two and pass on either side of the center.

This happens for two main reasons. The first is related to the heat island and the city breeze phenomena we've just met. The hot city creates a pillar of warm, unstable convective air. The wind is jostled by these columns of rising air, and if the heat island effect is strong enough, it is forced around the pillar—it is split.

The second reason is that tall buildings trip the wind. They turn a smooth-flowing wind into a turbulent, chaotic one. This can act as a barrier against the normal flow of the wind, which is then forced around the turbulence.

The two effects can team up and even split storm systems in two. Any weather that is split by a city often joins back together downwind of it.

THE MEANING OF BUILDINGS

The city shapes the winds, but the winds shape the city, too. Historically, the least desirable areas were always at the downwind end, where the poorer residents would have to endure the waft of industrial fumes and other "bad odors." The smoke has cleared, and most industry has moved to cheaper, less-populated regions, so it is less of a concern these days, but the weather still leaves its marks on buildings.

For many summers we spent a family holiday in a rural part of Brittany, where we had a small, basic lodging: a *gîte*. It had an oven, a kettle, some chairs, a table, and beds. There was no TV, no internet, not even a radio, and I loved it. It also had extraordinarily thick walls, small windows, and an enormous fireplace. I asked the locals about this architecture, and they explained that our gîte, which we found comfortable for two families, would once have been home to four or five peasant families. The summers were hot, the winters cold and damp. The thick walls and small windows kept the interior cool during the day in summer and retained some warmth at night, smoothing the temperature over the day–night cycle. Then, in winter, a fire would be kept going, and the thick walls would act as heat batteries, again maintaining a constant temperature.

That building looked a little unusual, but its architecture was a story about the climate. Every building in the world is trying to tell us something about local conditions. Architects who ignore the weather can be made to look foolish. In 2013 a new skyscraper in London, nicknamed the Walkie-Talkie because of its gentle concave shape, reflected and focused the sun's rays down on to the street below and melted parts of a car. Street parking was suspended and damages paid—and developers blushed.

At the extremes, the relationship between buildings and local weather screams out to us. Some of the buildings I once saw in Kiruna, in the Swedish Arctic, look like a dress rehearsal for colonizing the moon. In countries where heat is the challenge, breezes are treasured and the buildings reflect this. In parts of the Middle East, wind towers reach above the street to catch and channel any breezes down into the living areas. In cities like Hyderabad in India, wind towers and vents point into the wind. But most of the time the effect is more subtle, and in cities, noticing the clues requires lateral thinking and careful observation. You'll see more blinds and shutters on the south side of buildings, the same side that solar panels face.

In windy regions, architects tinker with the angles. The more sides a house has, the less vulnerable it is to strong winds; hexagonal or octagonal buildings stand a better chance against brutal gales. And roofs with several angles are more resistant than a simple up-and-down gable roof.

In many temperate regions, the weather gets a bit too warm for comfort in summer and too cold in winter. One simple architectural roof-design feature works to soften both. It is one of my favorites and we have it on our home. I'll give you a clue: The sun is higher in the sky in summer than it is in winter. If you were designing a building shape and you wanted to maximize indoor warmth from the sun in the winter and minimize it in the summer, how would you do that?

If a roof or other barrier is stretched over the windows below it has a double effect. It shields them from the high midday summer sun but still lets in the low winter sun. We have a desk at one of these windows, and it's easier to work at on a sunny summer's day, when it is in shade, than a sunny winter one, when the sunlight sneaks in under the parapet.

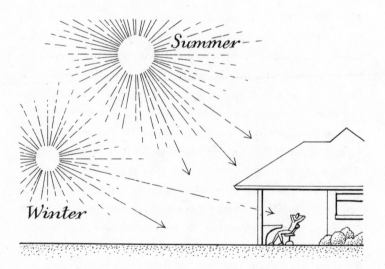

There is no shortage of flags on the buildings of Manhattan, and they joined the steam in mapping the lower winds for me. But in a city there are always flags that don't look like flags. Birds perching on the apex of a roof will tell you where the wind is coming from. If they are all facing one way, that's the direction the wind is coming from at that height. Birds like to face into wind because they need to take off into it to prevent it from ruffling their feathers. If the birds show little agreement in the matter and are facing in many different directions, that is a sign of light or variable breezes. This is common during high-pressure systems. If you notice the birds all facing one way and then, later in the day, they have turned, it is a sign that the wind has changed and the weather will soon, too. The same bird technique works on trees and in wilder environments, too, of course.

CHANGE THE LORE

More than half of the world's population now lives in cities. By some estimates, three million people a week move from rural to urban areas, and the proportion of city dwellers may reach two-thirds by the middle of the century.* This trend has some side effects, one being a burgeoning interest in city and suburban nature. We see some of the first flushes of this in the shift in weather lore, as it segues from watching wildflowers to noticing chairs squeak, supposedly a sign of humid air and worsening weather. Bed springs squeaking is a sign of dry air and good weather.

The problem is that no sooner has the lore caught up with our new lifestyles than we change them and our manufacturing methods. I can remember hearing chairs and beds squeak, but not in the past decade, so I can't vouch for these methods or even the likelihood of experiencing them. I still hear floorboards creak, and that is also supposed to be a sign of humidity, but I have not found it a reliable indicator.

---

* At the time of this writing, there are some early anecdotal signs that the COVID-19 pandemic may be reversing this trend. Time will tell.

Stone conducts heat well, which is why it feels cool to the touch, even on warm days. If humid air comes into contact with cold stone, the water vapor condenses and droplets form on the stone, hence the old adage, "When stones sweat, rain you'll get." It must also be why plates and soap have been seen to "sweat" before storms. (Although in the case of soap, I suspect a hot shower in the recent past was as likely a cause as a storm in the near future.)

It is said that mats swell, cheese softens, wounds itch, ropes tighten, and strings untwist in humid air. All of which may be true, but I have never found these useful. Anyone who has spent time on a boat or anywhere near water will be aware that salt definitely does clump and form chunks in response to humidity, but in my experience, it doesn't do it in a way that can give us a forecast. Windows and doors become harder to open and shut in damp air, and I've noticed this before a storm. We have a door that won't shut in certain conditions; cold and damp have the worst effect.

Perhaps the most arcane town-weather lore I've come across is the home run forecast. Balls, and all other projectiles including bullets, travel farther in humid air. It is counterintuitive, because muggy air feels somehow thicker to us and we know water is heavier than air. But humid air is actually less dense than dry. This means that baseball home runs, are, in theory, more likely just before bad weather arrives. And whatever your favored ball sport, you can try to draw some bizarre weather forecast from it. But don't race down to the bookie's: The humidity effect will not change the outcome of many games, because it increases the distances that balls travel by no more than 1.5 percent.

THE SIX WINDS OF THE CITY BRAVE

The clock was ticking. There was a limit on my solo exploration time, so I moved quickly, threading my way between buildings and along Manhattan avenues. I was on safari and keen to bag as many of the city winds as I could. This is not an expedition for the fainthearted: It takes a certain resolve to want to hunt them down.

By the time I met up with Ben and Sophie again, they were both draped in branded layers. We were all grinning. I had experienced seven of the sixteen winds, including the big six, and my wife and son had kept the wheels of commerce turning. We had to return to a sporting-goods shop to exchange a baseball mitt after I pointed out, with my limited baseball knowledge, that a right-hander might wear the mitt on the left hand. But, overall, both missions had been successful, and we strode off, keen to celebrate. The hot dog vendors were right to look afraid.

There is no point in my pretending that a detailed account of my scouring Midtown for these winds will entertain you, for this is not a spectator sport. Your satisfaction will come from seeking out these winds for yourself. It's a daunting task, and many will shy from it, but you've made it this far and I have faith that you'll prevail.

When a wind hits a tall building, it gives birth to sixteen other winds. Many are well above ground level and not easy to feel, unless you're a window cleaner. But we can sample six of them without equipment, and you'll recognize a couple. We met one earlier in this chapter and another on the downwind side of a tree.

We know that wind that hits the face of the tall building is deflected downward, so this is the first wind to feel: the downward rush of air we sense at ground level when standing on the windward side and near the building (1). We also know that this wind rolls into a vortex, and if we step away from the building, we should be able to detect the Monroe effect, an outward then upward waft of air (2). Walking a little farther away from the building, there is a strange calm area. In this spot, the wind traveling toward the building meets winds deflected in the opposite direction, partly canceling each other out, and we find an area of calm air (3).

If we walk to the corners of the building, we find that someone has turned up the wind knob, as the wind accelerates past the corners (4). Then, following the building around to the downwind corner, the hairdryer quiets and we experience weak, fluky breezes (5). Completing our circuit by walking around the downwind side of the building, we find the same double eddy we discovered downwind of the single tree (6). Stepping farther away from the building, we find that a wind blows toward it (7), appears to stop (8), then blows away from it again (9). The first six of these are the Big Six, the winds that the city wind hunter on safari will want to bag.

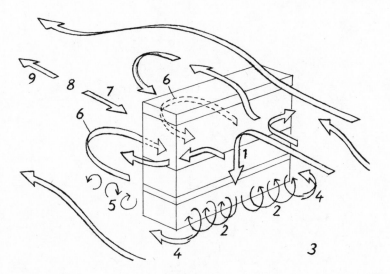

The only thing to add is that if we find ourselves on a roof-top, we can expect the wind to do all sorts of strange things. It is impossible to predict exactly what you will feel, as even the computers get confused about this, but the one guarantee is that the wind you feel will change dramatically with every few paces. If you're lucky enough to find yourself at a rooftop party, you'll notice something amusing, once you know it. There will always be comfortable and uncomfortable people on these occasions, but neither usually knows why. A knot of people will be putting on coats and lamenting how fresh and chilly it is, but barely ten yards away others will be taking layers off. The answer to a comfortable roof party lies not in clothing, but in moving.

Please do go in search of these winds. You'll enjoy the safari. I've found that it helps your chances of success greatly if you don't reveal the purpose of your mission. A few too many in our society consider their own follies superior to ours.

CHAPTER 18

The Coast

Seas and Seasons • Island Clouds • Base Colors • Coastline Winds •
Winter Sea Cumulus • Coastal Microclimates

THE PACIFIC ARCHIPELAGOS contain thousands of islands. The traditional way of life in regions like Melanesia, Micronesia, and Polynesia depended on interconnectivity between islands, which led to an intimate relationship with sea journeys and natural navigation.

Farmers know that a bad season will lead to a ruined crop, but the worst of it could be viewed, with grim resignation, from under a roof. In the Pacific, misreading the weather signs meant a poor harvest of fish if you were lucky, but being killed in minutes by a storm or after weeks lost at sea if you were not. It's unsurprising that these societies found meaning in shapes, colors, and sensations that passed all other cultures by. Reading weather signs blossomed into a life-or-death art form.

There are reports of forecasting methods that are too niche for us to want to adopt. On Nikunau, an atoll in Kiribati, navigators look for a coral that exudes a clear liquid in advance of settled weather and a milky one before storms. But the good news is that the main methods used in the Pacific are refinements of the disciplines we've been learning. This is refreshing, but shouldn't come as a shock: The culture, traditions, and way of life in the Pacific may be foreign to us, but the air, water, and heat are bound by the same rules.

Land warms more quickly than sea; islands heat up; the air above them rises and cools, vapor condenses, and cumulus clouds form. When warm, moist air encounters the rising ground of a volcanic island, it is lifted, and clouds form. The same processes we have seen over hills near home can be found above the places where people say, *Kia kohu te mata o Havaiki!* ("May the peaks of Havaiki be banked in clouds!")

The main difference lies in attention to detail. Most people would miss the clouds, so we might congratulate ourselves on spotting them, but did we look for the deeper meaning in their exact shape and colors? Even if we want to find these signs within signs, we must ask ourselves another question: Do we have the patience?

The Maldives share many physical geographical attributes with the Pacific Islands. Both contain at least a thousand coral islands lying near the equator, a long way from large land masses. The culture may be different, but the weather signs are the same. For the past couple of years I have done some teaching on the Maldives: I have run courses on reading these signs for locals, migrant workers, and tourists.

What struck me most during the sessions was the importance of time and patience, for me as much as for my students. Of course, the signs don't appear at a time of our choosing, but even when they do appear, they rarely announce themselves boldly. The colors, shapes, and motions that reveal so much are happy to remain in the background. We have to care enough to see what's before our eyes. Early on, it helps if you know how to force yourself to care.

Try this experiment. Look at a skyscape, then close your eyes and try to remember what you saw. Now look at the same sky-scape, knowing you're about to turn your back and list, out loud, every shape and color you could find. Imagine you're describing it for an artist who is facing the other way and hasn't seen it. If you have a willing partner to practice this with, the exercise is more honest and earnest, but if we're honest with ourselves we'll get there. And if I'm honest with you, I don't always find this easy. I had to rekindle this focus in the Maldives and was helped on the first day by having to stand at the end of a jetty for two hours, waiting to meet someone who was late. I recommend trapping yourself on a bench with a view for a couple of hours and taking the artist's view.

SEAS AND SEASONS

At one extreme, the relationship between land and oceans creates patterns that are, ironically, too big for us to spot easily. There are wind systems that last for months, and Pacific Islanders aren't the only ones who need to stay tuned. The word "monsoon" comes from the Arabic *mawsim*, meaning "sailing season." As the seasons change and the sun moves from large land masses to oceans, the pressure systems alter and ocean winds change direction.

In the Indian Ocean, during the age of sail, there were tra-ditionally two short seasons when commerce could take place: the *awwal al-kaws* and the *akhir al-kaws*, each lasting less than two months and governed by the monsoon winds. Those key wind seasons formed an integral part of coastal culture, and knowledge of their arrival was woven around the harvesting of dates and the start of the rains. To these cultures, the sign that rains were coming, the same clues we looked at earlier, meant that the en-tire region's weather was about to change—and for many weeks.

Ignorance of these seasonal changes would make any search for smaller signs a lot tougher.

In past centuries, any Westerners who visited the region and took to the seas without heeding local wisdom or understanding these shifts usually paid for their arrogance. In 1818, George Fitz-Clarence, Earl of Munster, conceded the point, but a little too late in his travels:

> Continual winds from the north-west, which blow nearly all the year round, have disappointed all my wishes and expectations. I am convinced the only good passages which can be made in this sea should be undertaken in November and December, and I might, if I had reflected before I came to the Red Sea, have concluded, from the scanty traffic on the outside of the shoals, that some good cause existed for its being so little frequented.

When it comes to patterns that are too large to spot easily, two approaches work. Spend a few years in the area and remain observant, or talk to the locals on day one.

Fortunately, on a smaller scale of time and distance, many signs are much easier to spot along coastlines and near islands.

ISLAND CLOUDS

Large volcanic islands shape clouds in much the same way that landlocked mountains do. There will be heaped cumulus clouds on the windward side and the lens clouds, flags, banners, and the other effects we saw downwind. They also split clouds in a way that happens around all mountains but appears more stark at sea. If you think of shallow water flowing past a pointed rock, it splits into two flows. The same thing happens to either side of tall peaks as clouds split into two whirling trains at either side of the island.

The effect on clouds above volcanic islands is sometimes bold enough to be seen from space.

Above smaller, lower islands we turn to the Pacific Islanders. The fine details that they learned to look for are also there for us to spot. Clouds that have formed above small islands as a result of local heating form certain patterns depending on the wind. We should not expect to see these effects much before midday, as the sun will not have had a chance to warm the land sufficiently, but from lunchtime onward, scour the horizon, study the relationship of land and sea, and gauge the wind.

If there is no wind, it's simple: Look for a cumulus cloud above an island and expect its highest point to be near the center of that island. If the island is big enough, with a mix of terrain, you may find the highest part of the cloud over a town or woodland, as the heat there is strongest.

If there is a light breeze, the cloud will be pushed along, much of it still over the island, but the top will be slanted in the direction of the wind. There will be no cloud over the windward side of the island, and most of the tallest parts of the cloud will be near the downwind edge or just over the sea in that direction. I often see some of the biggest clouds just downwind of the island, over the sea. This can lead to a stepped effect, with clouds that get gradually smaller toward the windward edge of the island until they disappear. The steps form in a line parallel to the wind.

Island clouds are constantly dissolving and re-forming, so one way to determine that you're looking at a cloud formed by an island is that it appears stationary relative to other clouds in the sky. It is very common to see a blue sky with fair-weather cumulus clouds scooting along with the breeze. The island cumulus will appear to stay rooted to the spot, or move very little. It looks like it's tethered to the ground, which in a sense it is.

From a distance, it is not just the static nature of the clouds that betrays them as island clouds but their appearance. Their size, shape, and brightness will differ from other cumulus clouds in the area. They are normally bigger, both wider and taller, and appear brighter.

There is an exception to the stationary rule. If a constant breeze is blowing and the island is big enough, the heating strong enough, and the air moist enough, then a train of clouds can grow large enough to survive even after they have moved away from the island. Once they have broken away, they stream downwind at the same speed as the other clouds in the sky, sometimes creating long parallel lines.

The rarest island cloud phenomenon I know of is a pair of "eyebrow clouds," known as *te nangkoto* to the Islanders. If there is no wind and very strong heating, a cumulus cloud will form. Then, if the thermals are powerful, they can break that cloud into two, forming a pair of eyebrows. I have seen split clouds over islands many times, which I suspected were a result of this phenomenon, but I have never seen any that resemble eyebrows. The search continues, and one day I hope to, but please do beat me to it!

BASE COLORS

Clouds over small islands are rarely rain bearing, which means that their bases are clearly defined and level. These flat bottoms can pick up a tint from the color of the surface below them. This is at the far reaches of most cloud spotters' abilities, and it takes a few elements to go our way. There has to be a stark contrast between two surface types, from a light turquoise sea to a dark wooded area, for example. Then you need just enough low clouds in the right part of the sky to make comparisons, but not so many that the sun's light is broken up: This shatters the light levels and makes subtle color shifts hard to pick up. The good news is that the air over the sea is humid, so the cloud base is usually low enough to be friendly for this exercise.

Looking for colors in cloud bases is an art that was born in Oceania but can be honed much closer to home. I have practiced it in many locations around the world, and I have come to expect both disappointments and pleasant surprises. The location where I have had the most practice is over a hill I can see from our home in the South Downs in Sussex. The top is cloaked in dense woods, making it higher and darker than the surrounding countryside. When low clouds cross this hill, they often take on a noticeably darker hue as they pass the top. I see it clearly about once a month, and by chance saw it this afternoon, as I was writing this chapter.

Keep in mind that the color difference is not the same as the land changing the clouds. We have seen how dark woods on hills lead to heating, which creates clouds and makes existing clouds darker, especially when the atmosphere is at all unstable. We are talking about something different here: the color reflections of land or sea in the clouds. Of course, the two effects can work in tandem and regularly do. To simplify things and to practice looking for color difference due to reflections only, I recommend watching low white stratus clouds pass over dark coniferous woods on hills. The blanket clouds mean that the hill and the local heating have less impact on their form and shade, which means that we can focus on the colors that are reflected from below. Look for a gentle bruise in the cloud, one that stays over the highest woods, even as the cloud continues on its way.

The challenge may feel daunting, but we're capable of sensing so much and there is a reason for every variation we see. It's a little like fishing: The pursuit is no less noble for failure, and we keep going because sometimes we'll bag something so wonderful that friends won't believe it, even the ones that let us tell them about it.

These patterns and techniques were honed by navigators searching for islands from the decks of their voyaging canoes. The cloud signs have to work, because the sailors' lives depended on their revealing hidden islands. But the joy for us is that we can borrow these techniques and practice them with the island in question in full view. I'm extremely fortunate that by walking up to Halnaker Windmill or onto the slopes of local hills, I can look down on a harbor with small islands and peninsulas across the water of the Solent toward the more substantial Isle of Wight. If I maintain a watch from early afternoon on a sunny summer's day, the south coast of England metamorphoses into the islands of Polynesia. But, unlike the Pacific Islanders, I have the training advantage of being able to draw a line from land I can see to the shapes and colors in the clouds. I encourage you to seize any chance you get to do the same. Whenever you have the opportunity on a calm, sunny day to look from a vantage point, with a little height, over one or more islands, you must grab it. Using the techniques above, you will see things that are in front of your eyes but happy to remain invisible for a lifetime.

COASTLINE WINDS

Earlier we met sea and land breezes, which are mostly steady and kind in nature. We also looked at how the sometimes powerful mountain and channel winds can follow a river valley all the way to the sea, charging out over coastal water from these mouths. There are two other coastal wind phenomena to add to our collection. They cause turbulent, sometimes violent conditions and are worth knowing about.

First, note whether a wind is blowing parallel or perpendicular to the coast. Next, look for any cliffs, scarps, or steep hillsides nearby. These landforms will trip onshore or offshore winds and send them somersaulting over the land or the sea just downwind of the feature.

This is the same phenomenon as the rotor winds we saw on the lee side of mountains, but the contrast at sea can feel greater. After an hour or two of kind breezes, the sudden arrival of strong gusts from apparently random directions is a jolt. The effect is compounded if the air is unstable, when it can act as the trigger to set a storm cloud growing.

If the wind is blowing parallel to the coastline, any jutting promontory or peninsula will create the same effect, only this time the whirlwind is horizontal, not vertical. Winds tend to hook back around any peninsula that points in the same direction that the wind is blowing. It is surprisingly common for an apparently sheltered bay to be hit by winds gusting in the "wrong" direction.

In physics terms, both of these effects are just eddies, and we have met them in different guises elsewhere. But near the coast, they cause more problems: Turbulence on a day of steady breezes is rarely expected and always unwelcome, especially if you're on the water.*

WINTER SEA CUMULUS

When I look down across the hills toward the sea, on many winter days I see lots of cumulus clouds over the water and next to none over the land. This is a reversal of the trend we expect to find at other times of the year. A land breeze can cause a line of clouds at sea, but if the clouds don't form a line that mirrors the coast, there must be a different cause.

The sea acts as a heat battery in early winter—there is a long lag in temperatures, which is why sea swimming in autumn is a better bet than in spring. On a cold autumn or winter day there

---

* Water patterns can be used to spot and track these winds before they hit you on a boat, but that is a different art, one that is featured in a book I wrote in 2016, *How to Read Water*.

may be far too little heating in the land for thermals, but plenty over the ocean. These clouds map the sea for us in the way that the island cumulus mapped land.

Once you're used to spotting them, you can see the sea in the sky. The cumulus clouds also show us that the air is less stable over the sea, so the winds will be gustier there, too.

COASTAL MICROCLIMATES

Seaweed that is hung outside absorbs moisture in the air and swells or grows slimy before bad weather. We can add that to our plant and animal collection and use it in a similar way; it enriches the picture but doesn't tell the whole story.

Most of the microclimate signs we have looked for so far can be used at the coast, but we need to keep in mind the powerful effect of salt. It can reach deep inland, but the 12 miles (20 km) nearest the coast are the most vulnerable. Winds carrying salt can have a devastating effect on plants and animals, and this dictates what survives in coastal microclimates.

A strange example of the power of salt can be seen when we try to read wind trends in coastal trees, a peculiar anomaly that natural navigators need to know about to guard against confusion. Near the coast, the winds from the sea can have a greater impact on plants than the prevailing winds. In Nova Scotia, for example, trees on cliffs have "flags" that point toward the west, even though the prevailing winds come from that direction. The trees have been bullied by the less common but very salty easterlies coming in from the sea.

Weather follows the same rules at the coast as elsewhere, but the contrasts are greater where land and sea meet. The weather speaks the same language, but the shapes, colors, and sensations are stronger and more intricate. Coastal weather characters are flamboyant.

CHAPTER 19

The Animals

Unweaving Lore, Literature, and Science • Dependable Signs •
The Birds • Insect Weather Maps • Insects and Wind •
A Richer Picture

T HE BIRDSONG STOPPED. There was some chinking from the blackbirds, and alarming from the tits. In the same minute that the sounds changed, the birds started flying from ground to trees. A pair of crows became vocal, and tussled. I watched a thousand tiny insects lift themselves from the wildflowers in the bright patch near my feet; they flew through the sunbeam and up into the canopy of the beech tree. The rain began minutes later. Two thousand years before that shower, Virgil wrote,

> Never has rain brought ill to men unwarned. Either, as it gathers, the skyey cranes flee before it in the valleys' depths; or the heifer looks up to heaven, and with open nostrils snuffs the breeze, or the twittering swallow flits round the pools, and in the mud the frogs croak their immemorial plaint. Often, too, the ant, wearing her narrow path, brings out her eggs from her inmost cells and a great rainbow drinks, and an

army of rooks, quitting their pasture in long array, clang with serried wings.

Like plants, animals react to the environment in many ways that help us understand the weather of the past, present, and future. Unlike the plants, they are mobile and vocal, which makes our task easier—most of the time. It also helps that animals are so finely tuned compared to us. Male adders take air temperature, sunlight, and wind into account before choosing a spot, then constantly tweak their body position to the sun's beams to maintain an external skin temperature as close to 93°F (34°C) as possible.

Many animals show a far greater sensitivity to humidity fluctuations than we are capable of, and we know that this is a dependable sign of change.

UNWEAVING LORE, LITERATURE, AND SCIENCE

If I'd been given a dollar for every time I've been asked whether cows lying down signal rain, I could have bought a calf. Sadly, little science backs it up, and I don't place any faith in it. Some say that cows lie down to keep a patch of grass dry, but I don't buy it. They're as likely to be lying down to chew the cud. Another theory is that cows lie down more in the afternoon, which is when most showers fall. But even if it's true, it suggests no connection with the rain and therefore has no forecasting value. But all cow watchers will have seen them congregate in sheltered zones, up against trees or in field corners, before and during gales, with or without rain.

There is a fascination with weather lore, but it poses a challenge: Should we embrace it, ignore it, or find a third way?

Let's start with a whirlwind tour of some of the "fun, but not proven" methods of yore. Before rain: Cats sneeze or get frisky, dogs wash behind their ears or eat grass, rabbits look in one direction, toads hurry to water, horses stretch out their necks, pigs grow restless, deer feed earlier than normal, the green woodpecker laughs, rooks fly crooked . . . the list could continue for pages, but we'll draw that tour to a close, fun though it is, and shift our focus to the animal signs that can be trusted. Lore is great for inspiration, visualization, and memory; science is better for true understanding. But these worlds are not exclusive, and in this chapter we'll be taking the best from both. Our dictum: Whatever works.

There are three types of dependable animal weather signs: individual behavior changes, group changes, and audible signs.

DEPENDABLE SIGNS

For a taste of the richness of animal signs available to us, let's zoom in on one particular animal behavior. An old piece of weather lore tells us that "Spiders' webs on long lines foretell a fine day, but if they shorten the threads it will rain." Can this really work?

Wind, temperature, and humidity affect how spiders spin webs. Temperature and humidity influence the architecture and nature of the silk in many ways, but the simplest truth for us to remember is that spiders spin smaller webs in windy conditions. This has been found to be true in academic research, including studies of spiders in wind tunnels. And we know that the wind often strengthens before rain, so, technically speaking, the lore holds true. We could argue that we can sense the wind without the spiders' help. But that is not really the point here. There is a direct scientific relationship between the size and shape of spiders' webs and the weather, and that is beautiful.

Natural navigators use the relationship between spiders' webs and the wind to find direction. Webs are more common on the sheltered lee side of barriers like trees, gates, or buildings, but there is an art to reading them. Spiders build in sheltered spots, but their webs also survive longer there. So a spider's web in isolation is a weak compass, while a collection of collapsed and new webs in one niche is a stronger sign. The "haunted house" look is proof that you're gazing at a spot where the wind doesn't reach to do any spring cleaning. This in turn indicates a nook that is protected from the prevailing winds, and in the US this is often the sheltered eastern side.

As you will have spotted, Thomas Hardy's books envelop the land and sky in a way that few others do. His characters refer to weather signs that can be found in toads, spiders, slugs, and sheep, and in one instance all four back each other up. Gabriel Oak, the young shepherd in *Far from the Madding Crowd*, sees a toad crossing a path, a slug that has crawled inside his home, and two spiders falling from the ceiling. Unsurprisingly, Oak looks to his sheep for confirmation and finds them "grouped in such a way that their tails, without a single exception, were toward that half of the horizon from which the storm threatened."

Prey animals habitually turn their backsides to the wind—it's as true of animals that weigh close to a ton as of a mountain hare. This is probably more comfortable for many species, but it also has survival benefits. Prey animals can see in a very wide arc, but not everything. Horses can see everything around them except the 10 degrees directly behind them, which is impressive, as it means their blind spot is only the width of an outstretched fist. The human field of view is only 120 degrees—our blind "spot" is more like two blind "landscapes": It's twice the size of what we can see.

If a prey animal puts its backside to the wind, it can pick up the scent of anything that tries to approach from the narrow sliver it can't see. A horse-whisperer I know told me that the wind changes a horse's behavior in another way: It is edgier in high winds, more volatile. Adam explained that strong winds agitate the scenery, and grass and trees sway, so it's harder for a horse to detect motion, making it feel more vulnerable and flighty.

Horses are rumored to sweat more before rain, and this rings true. Humidity levels rise before rain, and all sweating animals perspire more in humid air. We struggle to cool down when our sweat doesn't evaporate easily, so the body produces more. We put water on hot stones in saunas, not because it makes the room hotter but because it raises the humidity, which makes us sweat.

Herd animals can be seen as individual organisms, but also as a collective one. Flocks and herds range farther from "home," the farmstead, when the outlook is fine and draw closer when bad weather threatens. They also bunch together if there is a threat, including predators and bad weather, and spread out when relaxed. Any herd animals gathered in a clump in the shelter of a lee spot near the farm is a strong sign that bad weather is imminent. Sheep that have roamed and ranged over the hilltops indicate that they have sensed no dangerous animals or weather in the area.

Let's pick at some more of Gabriel Oak's notions. Slugs respond to changes in weather, notably temperature. Once on the move, they travel faster in warm temperatures than cool, but more interesting for weather watchers is how slugs react to *changes* in temperature. Slugs are typically nocturnal, as every gardener who dreads the dawn patrol will confirm, but there is one daytime situation that draws them out: a sudden chill, which is what we can expect before and during summer showers or cold fronts. Once the daytime temperature drops below 70°F (21°C), some slugs stir from under stones or plants, and their activity will vary in response to these fluctuations. If the temperature drops further, even more slugs become active, but when it rises again, they head home.

The males of many frog species croak after rainfall, as this is a good time for the female to lay eggs in fresh pools of water, but chances are that we've noticed that it's raining long before that. From January, if nighttime temperatures rise above 41°F (5°C), frogs and toads emerge from hibernation, but again, this tells us something we will already have felt.

Frogs grow more active with rising humidity, and this is more promising because it rises *before* rain falls. The most sensitive amphibian watchers will probably detect the changes, but the more casual frog appreciator will struggle. I've heard the frogs croaking during rain from Borneo to Bognor Regis, but I can't yet claim to have noticed them growing more active before a weather change.

You have probably seen earthworms at the surface after a rain. There is a popular belief that worms rise during rain because if they don't, they risk drowning in their tunnels. This is not true. Worms don't drown, they breathe through their skin and can survive for days waterlogged. They come to the surface after a rain, but scientists believe this is because it's a good time for them to travel longer distances—they can cover wet ground more easily than dry because they don't dehydrate on it. It's also possible that they surface if the raindrops sound like moles: They will surface to escape moles. Fishermen who like to use worms as bait have devised ways of simulating the sound and vibrations of rain to draw them up. If you would like to try this, one method includes drawing a saw blade over a stake in the ground.

THE BIRDS

The lore tells us that donkeys bray before bad weather, which is doubtless sometimes true, but any animal that is capable of voicing alarm can at times warn us bad weather is imminent. When it comes to audible weather signals, birds are the most trustworthy wildlife. The same small farmers in Botswana who were confident that plants offered clues were even surer of the animals' voices. Eighty-two percent agreed that "Through the chirpings of sóme birds and sounds from certain insects, I can predict whether it is going to rain or not."

We find the same confidence in rural communities across the world. The honeyeater bird in Papua New Guinea cries out loudly before rain. The chaffinch has a famous "rain call," often written as "huit," that it issues before rain—the call can be heard wherever the chaffinch is found, but amazingly, there are regional accents. I find that individual animal calls are not the key, but that any changes in the soundscape are definitely worth tuning in to. We'll return to this idea at the end of the chapter.

Birds help us to join the pieces. Simon Lee, a research meteorologist, once told me that he would watch red kites, a bird of prey, soar, orbiting on thermals. And he knew that the higher the birds went, the stronger the thermal must be. High birds meant very unstable air. Simon learned to use the height of the birds as a forecast: The higher the birds, the greater the chance of storms. In 2005 he saw unusually high birds over Yorkshire: Powerful storms broke not long afterward, and devastating flash floods swept in soon after that.

Woodland birds tend to forage lower down in the trees when the weather is fair and dry, and to rise up through the branches as it deteriorates. During fine spells, roosting birds stay out longer and head to their roost later. Both habits are reactive: The birds aren't giving us forecasts, but they are part of the jigsaw puzzle. On a personal note, I sense when animals like foxes and badgers are active at dusk, and this is closely related to bird roosting calls and weather, both present and recent. On hot, calm days, birds sound farther away than they actually are.

Birds tell us what the wind is doing as they fly, when they perch, and even when they sleep. One of the most common sights in the sky is birds gliding, but they struggle to do this in turbulent air. A bird gliding for long stretches is a sign that the air is stable, so this is a more common sight in settled fair conditions.

Many woodland birds, like pheasants, face into the wind when they roost. As we saw in chapter 17, urban birds likewise perch facing into the wind. Birds that hover, like kestrels, also face into the wind because they need to stay over the same spot on the ground: If they hover facing any other direction, they will be moved off it by the wind. A hover in this case is actually a very slow flight into wind. It is my favorite method of gauging the wind direction from a car: On any long journey in a noisy metal box, I find it satisfying to spot a wind vane hovering near the highway.

A specific type of bird hover maps the land for us. However graceful their flight, birds cannot defy gravity: If they want to climb, they need to flap their wings or hitch a lift on rising air. We have looked at birds circling on thermals, but they find one other type of rising air useful. Whenever wind hits steep ground, it is forced upward, as part of either an eddy or a wave. Birds are good at using this free energy, which is one reason you will often see them following the line of cliff edges or hovering above them.

Once we know this effect, we can spot it over a wide range of scales. In Pennsylvania, the prevailing wind hits the mountains and leaps up, and migrating hawks follow this line of rising air. But I have seen a similar effect above hedges, where birds appear to hover for a few seconds without flapping their wings. In each case they are pointing to a wind hitting a steep feature below them. I suspect that this effect is what Richard Jefferies was alluding to when he wrote, "The more I think, the more I am convinced that the buoyancy of air is very far greater than science admits."

Many birds fly closer to the ground when traveling against high winds, taking advantage of the slower wind speeds there. I regularly see birds, including crows, gulls, and pigeons, sweeping in fast and low over the fields when they're trying to make ground against a determined wind. They fly much higher, heading back in the opposite direction. Sometimes it's possible to see this effect in a single patrolling circuit. When looking for food, gulls or birds of prey fly over the nearby fields in a stretched oval pattern, a little like a running track. In strong winds, they fly lower on the upwind leg and a little higher on the downwind.

Birds sunbathe to warm up, and if they take off from a sunny spot, they're likely to be heading for another sunny spot. Sometimes they get too hot and move to the shade, even if this is within the same tree, but before they do this you may spot a bird like a jackdaw with its beak open, panting.

The tawny owls in my local woods definitely *seem* more vocal during fine, settled weather. "Seem" is important: Whenever we think we hear animals more clearly during good weather, we should ask ourselves, Are the owls really making more noise or has the calm weather quieted the trees, making them seem louder?

The limited science in this area suggests that each owl species responds in its own way. Some owls are more sensitive to temperature and humidity, others to moon phases and light levels. In the seventeenth century, Francis Bacon wrote, "The whooping of an owl was thought by the ancients to betoken a change of weather, from fair to wet, or from wet to fair. But with us an owl, when it whoops clearly and freely, generally shows fair weather, especially in winter."

When I hear owls hoot, I take it to mean they support my decision to go for a night walk. Bats often confirm it: They are more active in dry, warm conditions, when they find more insects to eat.

A pleasant recent obsession of mine has been to note the orientation of the woodpecker holes I see on my walks. The great spotted woodpecker, owner of most of the holes I come across, seems to favor the northeast side of the trees. It's not a hard-and-fast rule—there is one on the opposite side, barely a hundred yards from me—but it works when looking at the trend of many holes. Ventilation and warmth are important to woodpeckers, so sun and wind play a part, but the angle of tree branches is crucial. Woodpeckers won't tolerate water running into their hole after each downpour. One theory is that when prevailing winds bend the trees over toward the northeast, there's just enough of an overhanging angle in the branches to keep the woodpeckers' homes dry.

INSECT WEATHER MAPS

Butterflies grow more active and many species of ant accelerate as the temperature rises. Honeybees do not leave the hive to forage for food until it is 55°F (13°C) outside, and they grow more active as the mercury rises toward 66°F (19°C). A bumblebee that is basking and trying to warm up will press itself onto warm surfaces and align its body position to the sun. Some beetles raise their belly off the ground when it grows too hot, and others turn a white abdomen to the sun in the desert.

Temperature is critical to the insects, and they all react to it. Here's a good general rule: Over the course of a summer's day, you can expect to see large insects, then little ones, then large again. The bigger the insect, the better it copes with cool temperatures and the worse it handles heat. Big insects prefer the cool periods earlier and later in the day and will be less active in the middle. Small insects can't get going in the cool and have to wait until things warm up.

Midges cause a lot of discomfort in the Scottish Highlands, but their sensitivity to temperature can be used against them. They flourish in a narrow temperature band, neither too hot nor too cold, so if you're climbing or descending, the thermometer often rises or falls out of their comfort zone and shakes them off.

Crickets are famous for signaling a warm temperature, and the frequency of their sound is directly related to it. It varies with the species, but a typical reading on the cricket thermometer is one chirp per second at 55°F, increasing with the temperature.

Flying ants swarm in summer when it's warm and humid and the winds are light. They take to the air in such numbers that many believe there is a "flying ant day," when they come out together all over the country. But as we know, microclimates vary hugely over the land, so it should be no surprise that the single-day idea is a myth. There are series of days, when each territory's conditions are just right. If you see a swarm, they're telling you that it's more than 55°F and the winds are slower than 20 feet (6 m) per second.

According to some lore, ants traveling in a straight line means bad weather to come, but I have no observations or science to back that up. More interestingly, an Australian Aboriginal tradition has it that ants building a wall around their nest indicates heavy rains on the way, and this is echoed in Western weather lore. The Pacific Islanders we met earlier noted the way small red ants in the home would block their nests before bad weather and leave them open during fine spells; this, too, is found in Western lore. I haven't observed either, but it's hard to believe that both are without some merit if they crop up in such widespread cultures and in societies that had no contact at the time the beliefs developed. Further scientific research supports the idea that many ants are sensitive to humidity—weaver ants have been observed building nests before tropical storms.

In some species there is a strong relationship between the alignment of ant and termite mounds and the sun. In northern Australia, the compass termite (*Amitermes meridionalis*) is famous for building mounds that point north. They present a thin face to the midday sun and a much broader one to the early and late sun, which helps regulate the temperature inside their home.

There are more than thirteen thousand named ant species on the planet, and they each have their habits. We need to tune in to the behavior of our local species. I use only a couple of ant methods regularly, and they're closer to natural navigation than weather, but there is an overlap. Yellow meadow ants are common on well-drained grasslands in the UK and abundant on the chalk hills near me. They are a dull yellow in color, but their home is more helpful than the creature itself. Yellow meadow ants pile up soil to make nest mounds that are up to 1.5 feet (0.5 m) high. The mounds are rounded bumps, but they are not perfectly round: They have a flatter side, and this normally faces southeast, acting as a solar panel and harvesting the sun's warmth on cool mornings.

The ant mounds are a great demonstration of the way microclimates vary over tiny distances; every anthill in the world will reveal something about their own surrounding microclimate. A few years ago, I was exploring Petworth Park in West Sussex with my friend the weather expert and former BBC forecaster Peter Gibbs. On this occasion we were meeting for work, as I was a guest on BBC Radio 4's *Gardener's Question Time*, which he was hosting. In truth, I was a little nervous about appearing on the show. It's something of an institution in the UK, with a cult status and a large, devoted audience. My heart beat faster as I waited for Peter and the crew to arrive at our meeting point in the car park.

We were soon underway exploring the park. I pointed out a few things, and then we stumbled on something worth investigating more thoroughly. We spent a few minutes looking at how the wildflowers were different on the north and south sides of the yellow meadow ant mounds. The light-loving flowers, especially wild thyme, were reveling in the brighter, warmer southern side of the mounds. The colors made it clear that there were very different climates on either side of the hills that didn't even reach our knees. The ants are just as sensitive to it. I had such fun looking at these little clues that I was lost in our own microworld and totally forgot we were sharing it with a couple of million listeners.

Morning is a good time for studying butterflies, as they lie still for longer in the cool and flit more when the warmth of the afternoon arrives. Butterflies are more sensitive to moderate changes in temperature than to wind, but they behave in almost the opposite way to slugs. Both are cold-blooded and need a certain temperature to get going, but butterflies retreat when the temperature drops.

Butterflies are also sensitive to the sun's radiation and are more likely to start flying when they are in direct sunlight. When sunlight is in short supply, some butterflies will fight over it, and you may spot them battling over sun flecks in shady woods. (Some dragonflies also scrap over them.) The butterfly's sensitivity to sunlight can be seen in the way it holds its wings: If it's holding them up, it's probably resting; if they're open, it's basking in the sun, trying to warm itself. If a butterfly with open wings takes off from a sunny spot as you approach, it will most likely land in another sunny spot. If it takes off from shade, it will probably land in shade.

(A few weeks ago, I accidentally herded a peacock butterfly down a Sussex farmer's track on a warm autumn afternoon. Each time I drew near, it took off from its sunbathing, flew a few yards down the track, and alighted on a new sunny spot, opening its wings directly to the sun. I took a photo of the butterfly and used it as part of a quiz for my email newsletter: Which way was I looking when I took the photo? It was afternoon, so the sun was in the southwest, the butterfly was opening its wings to harvest this radiating energy, so I had to be looking opposite the sun. The answer was northeast. These are the sorts of puzzles that natural navigators enjoy when we're not able to immerse ourselves in nature.)

Butterflies are less active in cloudy conditions, flying over shorter distances and times. Rain is a real problem for them, so it makes sense that we see less of them when the weather cools or grows cloudy. I remember this with a little rhyme: "Butterflies are shy of the clouds in the sky."

INSECTS AND WIND

Mary Schäffer was one of the first nonindigenous women to head deep into the Canadian Rocky Mountains. Not many nonindigenous men had been there either. She was an explorer, a pioneer, and an expert with horses. In September, her party descended from the higher valleys to the Kootenai Plains. It was warmer lower down and they were plagued by millions of sand-flies. They knew that the flies were thriving in the "soft chinook winds," the warm "snow-eating" foehn winds of North America. In her account, first published in 1911, Mary wrote, "Next morning the chinook wind was gone and with it the flies."

Flying insects are very sensitive to wind. Even midges that are happy with the temperature will be sent packing by a Force 3 wind, which is only a gentle breeze. On a windy day, if your path takes you into the lee of a woodland or the still air behind a hedge or rocky outcrop, you may feel relief at the calm, followed by surprise to be joined by so many tiny airborne creatures. If you're lucky, some pragmatic butterflies may join the party. If you then walk into a wood, you may notice that the insects change in size and appearance very suddenly. The bigger insects, like dragonflies, are strong enough to brave the winds outside the trees, but in the wood, we find the weaker tiny insects. Amazingly, there is some evidence that on windy days, biting insects, including mosquitoes, bite prey on the lee side. Something to think about as we put on the insect repellent.

Old lore tells us that a bee was never caught in a shower. I have seen bees in the early stages of showers, but I can't recall seeing one flying in heavy rain—I'd feel for it if I did. Beekeepers confirm that honeybees are sensitive to rain and don't swarm if bad weather looms. Studies have also shown what we'd guess to be true: Honeybees visit flowers less often in low temperatures, heavy rain, or high wind. But much more exciting, the results show that the visits dropped significantly when humidity rose, too. The bees sense and react to temperature and humidity changes, so their behavior can warn of bad weather approaching.

There is an intriguing and mysterious small flying insect known variously as a thrip, harvest bug, storm bug, or thunder fly. There are about 150 species in the UK and more than 700 across the US and Canada, and most are less than 2 millimeters long. There have been reports since the mid-nineteenth century that these insects appear in the thousands shortly before a thunderstorm.

Like bees, thrips are most likely to take off in great numbers during warm summer weather. And as we know, thunderstorms are more likely in these conditions, too. The insects swarm when the conditions are right, but that is not what they're famous for. Country people have noted that these insects stop flying and land in vast numbers in certain places just before very bad weather. The fascinating question is: What triggers the insects to land before the storm? One theory is rising humidity but, much more intriguing, they may be sensing changes in the electrical fields in the atmosphere. One day the mystery will be solved, but we're not there yet.

A RICHER PICTURE

The birds and insects are busy mapping the microclimates for us, adding their own layers, which sometimes overlap, something Richard Jefferies noted when watching swallows: "If not wheeling in the sky, look for them over the water, the river, or great ponds; if not there, look along the moist fields or shady woodland meadows. They vary their haunts with the state of the atmosphere, which causes insects to be more numerous in one place at one time, and presently in another."

Like the plants, the animals may not offer us the first signs of change out there—Gabriel Oak noted the animal signs after spotting the "sinister aspect" of the sky—but they can remind us with a nudge to tune in to the sky and other signs. This is often how I use them. Perhaps I have been in a woodland for a while and have lost sensitivity to the sky. If I notice changes in animal behavior, or the woodland sounds alter and the cause is not obvious, I'll search for the trigger. Has the wind shifted? Can I see changes in the sky through a canopy? This leads to a less romantic but more practical and realistic use of animal signs. The animals are not performing stand-alone magic tricks, but they are part of our awareness clothing. This perspective adds richness to our understanding, attention, and appreciation. The holistic approach sounds ethereal, but let me show you this in practice.

I can see a looping long branch of a spruce tree from my cabin window. A pair of buzzards patrol the area, and this branch is one of their favorite perches. The songbirds and crows are also happy to use it, if the buzzards are away on maneuvers. The shape of that spruce and the shorter trees and plants around it have been sculpted by the prevailing southwesterly winds. We know that birds perch facing into wind, in the country as on city roofs, so this means that the perching birds face southwest more commonly than any other direction in my part of the world. To my eye, the direction the birds face "fits" with the shape of the plants nearby. The patterns chime.

Every morning I try to take in the skyscape, the upper and lower clouds, and the winds that are carrying them, as well as any dew, frost, mist, or other signs. But I don't always take the time to spot everything out there. Later in the day, if I notice a bird—or better, two or more birds—facing a direction other than the typical one or the one I saw earlier in the morning, it sets the wheels of curiosity spinning. I scan and scour, and this is so often the moment at which I pick up cirrus or other signs that have probably been there for an hour or more without my noticing. The birds have pecked me on the shoulder and whispered, "Stop daydreaming and start sky watching. Change is coming."

CHAPTER 20

Storms

The Top and Bottom of a Storm Cloud • Storm Systems • Tornadoes •
Hurricanes

JULY 2019 BROUGHT A HEAT that started in the air but soon seized the ground, the water in the pond, and then our bones, until bedtime brought only thoughts of restlessness. Later that week, Cambridge Botanic Garden saw a temperature of 101.7°F (38.7°C), a new UK record.

A couple of weeks later, military helicopters dropped tons of aggregate in an attempt to defend the village of Whaley Bridge in the Peak District against the surging floodwaters. The dam was damaged and at risk of bursting. Police ordered residents to evacuate, and Whaley Bridge became a ghost town.

Heat waves rarely step quietly off the stage. They like to leave with bangs, rumbles, and torrential rain. Summer heat can be thought of as a rubber band: It will stretch so far, but there comes a point when it snaps. The snapping point is a thunderstorm, and it's governed by our old friends heat, water, and stability. At any one time, about two thousand thunderstorms are raging around the world.

A question I'm asked regularly before someone books an outdoor course is: Will it be canceled if the weather forecast is bad? After twelve years of running them, I can't recall canceling because of the weather. Rain, snow, hail, sleet, and wind are not course-cancelers in southern England. In these conditions I like to joke that students get double the value: It feels like you've been on the course for twice as long, but there's no extra charge.

I have changed the timings of courses and altered routes, usually because of the risk of a thunderstorm. I'm not blasé about storms, even in the gentle hills of Sussex. I will never lead a group anywhere near the top of even a small hill if there is a serious risk of lightning. We can all gauge the risk more effectively by studying the anatomy of one particular cloud, the cumulonimbus, and that is much easier once we know how to spot an important layer in the atmosphere.

We expect the air to get colder if we climb mountains, and it does. The air gets drier and cooler with altitude, losing on average about 12°F (6.5°C) per 1,000 yards. Many people might guess that this trend continues indefinitely, the air growing colder toward the tops of the tallest mountains and beyond, all the way to space. But it doesn't. It gets colder all the way to the highest summits and a little higher, but then something odd happens.

In 1899 the French meteorologist Léon Teisserenc de Bort discovered a small detail with big ramifications. About 6 miles (10 km) above Earth's surface the air temperature stopped decreasing with altitude and actually started going up. De Bort had revealed one of the most significant boundaries in our atmosphere, one we now know as the tropopause.

Above the tropopause, we are in the stratosphere, where the rules change and the air starts getting warmer with altitude. Bizarrely, this means it can be roughly the same temperature at an altitude of roughly 30 miles (50 km) as it is at sea level on a winter's day. It's odd to think that it's warmer six times higher than the summit of Mount Everest than it is on top of the mountain, a thought that shouldn't distract you from the fact that without pressurized breathing apparatus and astronaut-worthy clothing, you'd be dead.

The cumulonimbus, the storm cloud, forms in the same way as other cumulus clouds, the only difference being that the processes are in runaway mode. As the vapor condenses to liquid water, more heat is released than is being lost as the warm air expands. The cloud keeps rising until it hits a barrier at the tropopause, a mighty temperature inversion, a triple-glazed glass ceiling. The height of the tropopause varies, from about 30,000 feet (9 km) over the poles to 56,000 feet (17 km) nearer the equator, which means that some cumulonimbus clouds dwarf the highest mountains and can be seen from nearly 200 miles (300 km) away. For most of us, 36,000 feet (11 km) is a good estimate.

It's possible to gauge the rough distance of a mature storm cloud—that is, one with a flat top or anvil shape. If you stretch out your fist on its side, with the thumb on top and your little finger nearest the ground, that will make an angle of about 10 degrees. If the cloud is the same height as your fist, it is a little more than 160,000 feet (50 km) away. If your fist only reaches halfway up the cloud, it's half that distance, just over 80,000 feet (25 km) away. This works best if you can estimate sea level: Place the bottom of your fist where the horizon would be if you were at sea.

If you see a storm cloud in the distance, try to note whether it is moving from left to right or right to left. If either is true, that storm is unlikely to pass over you. Only storms that maintain a constant bearing are heading your way. (This is one use of a technique I learned as a sailor and pilot: It can be applied in a wide variety of outdoor situations, and is best summed up as: "Constant bearing, constant danger." If something is going to collide

This storm is about 31 miles (50 km) away and moving from right to left.

with you—boat, airplane, storm, bumper car—then its bearing, or its angle relative to you, doesn't change. From the cockpit of a small aircraft, another aircraft on collision course with you can appear like a fly stuck to the windshield—it doesn't move, just grows slowly, then terrifyingly quickly.)

A typical storm cloud's life cycle is only about ninety minutes from start to finish,* so most of the thunderstorms we see in the distance will never reach us. Distant, isolated storms offer a great opportunity for us to watch their life cycle develop from a comfortable and safe position.

A good starting point is to look for cumulus clouds that have grown much taller than they are wide. If we keep a close eye on them, we can catch the moment that they turn into something more sinister. Study their tops and look for definition: Can you

---

* There are a few exceptions, like the storm complexes that develop in places like the Great Plains, in the American Midwest.

see the rounded tops, like cauliflower florets? This shape is tell-ing us two things: first, that the cloud is still rising and second, that the highest part of the cloud is still liquid water, which is a lot more important than it sounds. Keep watching the top, and if the cloud continues to rise there will come a moment when the top changes in appearance, becoming less defined and much fluffier and wispier, like cotton candy. You may recall these de-scriptions from cirrus clouds, and that is no coincidence: They are both signs of ice forming. Once the top of a cumulus cloud turns to ice, it's well on its way to being transformed from an energetic cumulus cloud into a potentially destructive cumulo-nimbus.

Keep watching the cloud develop and you will see it smash up against the glass ceiling of the tropopause and start to spread under this temperature inversion. This gives the cumulonimbus a flat-top anvil shape: It has reached middle age. Imagine pouring a thick batter from a bowl onto a kitchen counter: It flows fast vertically, then stops and spreads out horizontally. Storm clouds do the same, only upward.

Soon the cloud's power will fade, and it begins to sink down to more of a collapsed shape. The recently mighty form dissipates and crumples, sometimes leaving a collection of cirrus and high cumulus clouds in its wake.*

Every thunderstorm cloud will display the three stages of life: growing, mature, and dissipating. The first two stages take place rapidly: This is when we need to be the most vigilant. A cloud can easily turn from a tall, friendly-looking cumulus to a mature and raging storm cloud in half an hour. The heaviest precipita-tion, rain or hail, falls from the mature cloud. The decay stage is

---

* One particular type of cloud is commonly seen after storms. It has been given a new name, "Asperitas," by Gavin Pretor Pinney, founder of the Cloud Appreciation Society. It has a rough, wavy formation and is commonly seen over the Great Plains.

more leisurely; it may take more than an hour for it to crumple completely.

The same rule applies as with all precipitation: A sudden start means it's soon over, and rain from storm clouds rarely lasts longer than half an hour, often much less. There may be more than one of those clouds out there, but the heaviest rain will not last long in each burst.

Many people believe that storm clouds return, but they never do. A new storm cloud can and is quite likely to form in the same spot as an earlier one, for all the reasons we looked at in earlier chapters that cause convection. Even knowing this, you may feel as if the same cloud is growing vengeful, especially when you're in a boat or a tent. But it's always because you're in a favorable spot for these clouds to develop and a new one has done so. Storm clouds are copycats.

If we see storm clouds or suspect that storms are in the area, we need to figure out if they are isolated or part of a frontal system: Do they hunt alone or as a pack? Storm clouds caused by local heating are isolated, and we can try to spot the trigger in the landscape, as we did with the more modest cumulus clouds.

If the air is warm, moist and unstable enough, it takes only a little jumpstart to set off the chain reaction. It may be a breeze carrying this air up a hillside or over a coastal cliff; it may even be the sun heating the woods more intensively than the neighboring farmland. If the conditions are primed and ripe for a storm, the trigger can be surprisingly modest. If high ground is the cause, the storms are more likely on the windward side.

In 1975 a particularly impressive local storm dumped three months' worth of rain on a patch of north London in a few hours, leading to shock, awe, and the cancelation of the BBC Proms concerts at the Royal Albert Hall. The area affected was tiny, but the intensity of the storm was so great that an investigation was launched to understand what had caused this very local tantrum of the skies. The meteorologists thought the blame lay with a hill in Hampstead, but the highest point in the area is barely 450 feet (137 km) above sea level, and only 250 feet (76 km) above the surrounding land. It is not a mighty hill, but the energy wasn't in the ground: It was in the air. It takes only a small match to light a big firework.

Isolated storms are most common in the warmest summer months, peaking from June to August, leading to the old joke, attributed to George II among others, that a British summer consists of "two fine days and a thunderstorm." Summer storms nearly always depend on the sun's heating of the land, so they are much more common in the afternoon than in the morning and tend to die away after sunset.

I think we all have an inherent sense of when summer storms are growing likely—maybe we retain this knack as part of our evolutionary survival toolkit. People talk of the weather feeling "close" during a period of extreme mugginess, and you may hear someone say, "The weather is going to break." These are all valid sensations related to heat and humidity, but we can add a layer by keeping an eye on visibility. If visibility deteriorates and the air grows hazier on such a summer afternoon, this is another sign that we're getting near the snapping point and storms are likely within hours. As we did earlier, try to identify a distant object and monitor how easy it is to pick out details: Have you lost sight of the windows in a house or the individual branches of a tree?

In winter, cold air dominates and the conditions don't seem ideal for thunderstorms, yet we experience plenty. Winter storms owe less to local heating and more to frontal systems, cold fronts in particular.

A cold front drives a steep wedge of cold air under the warmer air ahead and lifts it, which often leads to storms. Instead of the air being blown over a hill, it's as if a steep mountain is ramming itself under the warm, moist air. And because fronts are broader than hills, this leads to a band of storms marching over an area, a little like battlefronts.

An early-warning sign of thunderstorms is "castles in the sky." If you see dozens or even hundreds of cumulus clouds shaped like towers or turrets—formal name *Altocumulus castellanus*—they guarantee a wide blanket of moist, unstable air. Frontal thunderstorms are likely within hours.

If you hear several thunderstorms in the middle of the night, we can assume that a cold front is going through and the air will be cool and fresh in the morning, with great visibility.

Here is an intriguing piece of old weather lore: "A thunderstorm in April is the end of the hoar frost."

This is interesting because, like so much lore, it hints at a truth by connecting things that have no direct scientific relationship. There is no causal link between thunderstorms and frost, but it is quite possible that a thunderstorm in April will mark the end of hoar frosts until autumn. Take a moment before reading on to see if you can figure out which type of storm would be most relevant here—isolated or frontal—and why.

If we see an isolated thunderstorm in April it's a sign that the air is very warm and the sun has given the land enough energy for a localized cumulonimbus to form. Both indicate that a warm air mass has arrived and the sun is now warming the land strongly as well. A frost after this is not impossible—the sudden arrival of a cold air mass might do it—but given the right conditions, a warm air mass and sunny period in April could well see you through the last of the frost season.

THE TOP AND BOTTOM OF A STORM CLOUD

Inside storm clouds there are always very strong updrafts and downdrafts. It is these currents, these violent elevators, that make hail possible, but they also leave footprints in each cloud that we can learn to read. In the following examples, we'll assume that you have spotted an isolated thunderstorm from a safe distance, allowing you a chance to study its anatomy. Later in the chapter we will tackle the times when we aren't so fortunate to have a safe view.

When a cumulus is on its way to becoming a cumulonimbus, the updrafts are dominant and powerful: They can be traveling at 30 mph (48 km/h) vertically upward. Sometimes they are so punchy that they can lift the air above the cloud, causing this to cool and condense into a cap cloud above the main cumulus cloud. This cap cloud is known as a pileus. If you see a cap above a tall cumulus cloud, it is a strong sign that a storm cloud is forming.

Once a storm cloud has formed, there are still very strong updrafts, which can push at the top of the main cloud. If an anvil has formed, there may be a bulge poking just above the anvil. The bulge also indicates extraordinarily strong updrafts and is a sign of a powerful storm.

If you look closely at the anvil, you will see that it is not symmetrical: It is shorter on one side and more drawn out on the other. The stretched side is showing you the direction of the winds at that level, and this is the direction that the cloud is traveling. The anvil is a fat finger pointing the way the storm is moving.

If there is no clear finger, it may be your perspective, or that the upper winds are weak. Try checking the direction of any other high clouds, like cirrus. Now that you have determined the direction that the storm is moving, you may also notice that the main cumulonimbus cloud looks slightly different on each side, too. The air isn't moving up and down equally in all parts: The updrafts are most powerful on the upwind side, which gives it a steeper appearance. It is also why, if you see a bulge above the anvil, it will be just this side of the center. On the downwind side, the side the finger is pointing, the top of the cloud appears flatter and more spread out.

The same forces that can cause a bulge at the top of storm clouds are at work at the bottom, where downdrafts may cause hanging bulges, called mamma. These pendulous protrusions are much more common in the later stages of a cumulonimbus, when downdrafts have started to dominate over the updrafts. They are a sign of a vigorous, energetic cloud, but one that has passed its peak.

Occasionally you may spot a funnel descending from the base of storm clouds. These clouds, known formally as tuba, are a sign of active drafts within the cloud. The rotating funnels often don't reach much below the base of the cloud, and they disappear once it has lost its power. However, in certain situations, not least in parts of the US, the funnel may grow and reach toward the ground. This can be an early visual warning of a tornado developing.

Sometimes a cloud rides ahead and heralds the imminent arrival of the main storm cloud. It is called a shelf or arcus cloud because it may appear as a flat shelf or wedge spreading out from the base of the cumulonimbus. The shelf cloud forms when downdrafts spread after they hit the ground, then lift the warmer surrounding air. If you turn on the kitchen faucet you will see the water flowing downward, then spreading horizontally over the flat surface of the sink. The main storm turns on the wind faucet, and the shelf cloud is carried out by the spreading winds at ground level. Unfortunately, it doesn't give you much warning, as it is attached to the main cumulonimbus. If a shelf cloud passes over you, the chances are that you will be more focused on the gusting wind than on the clouds above.

In the examples above we've been sitting in a comfortable movie theater, grabbing fistfuls of popcorn and watching an isolated storm cloud grow, mature, and decline from a safe distance. Experience tells us that this is not what we can expect every time: Storm clouds like to creep up on us, cloaked in other clouds or hidden behind them.

In 1938 five glider pilots were searching for rising air during a competition when they flew into a friendly-looking bank of clouds. They found the lift they sought, but far too much of it. Unfortunately, a cumulonimbus was lurking inside the other clouds and the gliders were torn apart. There was one survivor.

The good news is that, just like bad guys in movies, the storm clouds that try to creep up on us at ground level can't help giving themselves away. In films, the clue is that the music changes; in the outdoors, the wind does.

In 1985 a Delta Airlines aircraft was approaching the runway to land at Dallas/Fort Worth airport, when the pilots saw lightning ahead. The plane continued on its path and flew into a thunderstorm. Seconds later a vigorous downdraft of air, a "microburst," pushed the plane violently down. The pilots tried to fight it, but it was no good. They tried to abort the landing but it was too late. Flight 191 crashed short of the runway, killing 137 people, including a twenty-eight-year-old driving his car on the freeway next to the airport.

Thunderstorms are a story about violent vertical air motions, but the kitchen faucet reminds us that the winds don't stop when they hit the deck. They lead to gusting horizontal winds at ground level, too. Cold blasts are common around storm clouds because some air is falling fast from high altitudes, and also because the rain cools the air below the cloud. If you ever feel a sharp, much colder gust of strong wind that you were not expecting, it's a storm warning.

We feel gusts of wind all the time, thanks to the eddies, waves, and other ways in which wind flows past obstacles, but the combination of a sudden change in wind speed and a drop in temperature is an alarm bell. If you can't spot the cause—a sea breeze, for example, or a katabatic wind rolling down off a snowy hill—then it's time to heighten the senses and weigh your plans. This is the moment in the film when the jarring violins kick in and the young woman decides it's a good idea to go into the deserted building at night on her own. You might choose to head off on a solo expedition at that moment, too, but if you do, make sure you read to the end of this chapter.

Another widespread misconception about thunderstorms is that they can travel against the wind. Once we understand how cumulonimbus clouds change the winds all around them, it's easy to see how this misunderstanding can spread. And I nearly learned this the hard way.

I was sailing off the Isle of Wight in 2007 when something odd happened. I was doing some preparations for my solo transatlantic sail, and on this occasion my main task was to come to grips with a bizarre and wonderful piece of equipment called a self-steering wind vane. By using the power of the water flowing beneath the boat and the angle of the wind, the wind vane uses pulleys to adjust the tiller and steer the boat. It's genius. It can hold a boat on the same course relative to the wind for a thousand miles—with no batteries, electricity, or external power of any kind, other than the water flowing beneath the boat and the wind.

The whole point of this mechanical marvel is that it keeps the boat steering in the same direction *relative to the wind*—it doesn't care which way the wind is blowing. If you set it up so that the wind is coming from the starboard beam, you can leave it for hours or weeks, and the boat will still have the wind on the starboard beam. I could rattle on happily about this invention, the savior of single-handed sailors the world over, but that is not what we're here for, which is to understand storms better.

That day I was new to using this equipment. I understood the theory, but I'd spent only a few hours using it when the boat started turning and didn't stop until it had completed a full U-turn that I definitely wasn't expecting. This was alarming. When I checked the wind vane, it was functioning well and I was still sailing with the wind coming over the starboard beam. Everything worked perfectly, but I was now sailing in the opposite direction to the one I had expected. The wind had performed a U-turn and the system had followed it. This 180-degree turn definitely pointed to something strange going on, but it wasn't in the boat. It was in the sky.

Scanning the horizon anxiously, I suddenly spotted the piece that solved the puzzle. There was a growing cumulonimbus not far off. It had snuck upward as I focused on ropes, pulleys, and levers, and it was too close for comfort. The wind was now blowing directly toward the fast-climbing cloud. I reefed the sails right in, disengaged the wind vane, took hold of the tiller, and raced for home on growing winds and adrenaline.

Thunderstorms move with the wind, in the direction of the higher wind, as indicated by the anvil and any cirrus clouds, but it is common to find yourself facing a storm cloud as you watch it approaching, and to feel a wind blowing from behind you toward that cloud at the same time. The wind you're feeling is the air being sucked into the growing cloud from below. Imagine the kitchen faucet working backward and sucking the water out of the bottom of the sink. The effect is especially strong in the first half of the cloud's life, as it grows to maturity, because this is when updrafts prevail. Once the cloud is mature or collapsing, downdrafts are more common and gusts that blow away from the base of the cloud are more likely. A storm cloud moving "against the wind" is a sign that it is young and still growing.

If you do find yourself looking at the base of a cumulonimbus, you may notice differences in shade, texture, and color. The darker regions are the updrafts and the lighter, rougher regions are where downdrafts and heavy rain are.

There are several theories behind the root cause of "the calm before the storm." The first thing to note is that everything, weather or life in general, seems relatively calm in hindsight after a storm hits us. When it comes to the weather, the theory I favor is that there are two different types of calm before a storm.

In the first, an isolated storm on a warm summer's day, the phrase refers just to the contrast between a bucolic summer's day and the mayhem of a storm. But the second type is more intriguing. When there are frontal storms, things are less calm before the storm starts—there are winds already. But the storm cloud changes the wind at ground level, and this can nullify any prevailing wind just before it reaches you. So for an hour you may feel a wind blowing that disappears as it is counteracted by the approaching cumulonimbus, just before you feel the full effects of the storm.

The average storm cloud has more energy than an atomic bomb, and we get a taste of this energy in the ferocious winds it brings, but it's the thunder and lightning that really quicken the pulse. Lightning heats the air to approximately 54,000°F (30,000°C), six times hotter than the surface of the sun. The air expands forcefully, sending out a shock wave of sound, which we hear as thunder. Light travels much faster than sound, which is why the gap between the two allows us to gauge the distance of the lightning. For every three seconds that pass between the flash and the thunder, the strike was about half a mile away.

Lightning lasts a fraction of a second, but thunder can rumble on for much longer, which is a bit odd, since one causes the other. This happens because the sound is coming from more than one place. Lightning strikes within one cloud or from one cloud to another are more much common than lightning from cloud to ground. Even the lightning of ground strikes may be a mile or more long, so it takes the sound from the highest part of the lightning seconds longer to reach you than it does from the bottom part.

Ground-strike lightning makes a round trip from cloud to ground and back, and the main lightning we see and hear is the "return stroke." Like most people, I see the lightning travel down to the ground, but this is a good example of our brain rewriting and simplifying the story to suit itself. There are stepped leaders of ionized channels of air, or "feelers," that we can see as spidery arms of light that reach down from the cloud. They move more slowly than the main bolt of lightning and branch regularly, which is why we think of lightning forks. They are known as stepped leaders or feelers because once one gets near enough to the surface to complete the circuit, the return stroke kicks in: a much straighter, much more violent bolt of lightning. It is the one we hear and feel.

You might see flickering lightning, and it's tempting to think of this as light bouncing off clouds, but light moves too quickly for our eyes to see it as a flicker. The flickering is multiple return strokes: The current is pulsing from ground to cloud.

Cloud-to-cloud or cloud-to-ground strikes cover the vast majority of lightning we will experience in a lifetime. You may hear different expressions used to refer to lightning that make it sound as if there are many more types out there than there are. "Heat lightning" just means it is too far away to hear or see clearly: It causes the weak flashing glow in the sky we sometimes see before any storms reach us. The lightning is no different, it's just far away. "Sheet lightning" is a term used when we see a broad white light, rather than a flash, and it just means that the lightning is totally enveloped in cloud. We don't see the stroke, just the cloud lighting up.

You might see "anvil crawlers," which are a type of cloud-to-cloud lightning. As the name suggests, these are spidery tendrils of lightning that reach sideways from under the anvil. They can be impressive and reach a length of up to sixty yards.

Ball lightning is different, and a real but very rare phenomenon. It seems to be caused by a normal lightning strike that somehow creates a supercharged plasma sphere. The words "seems" and "somehow" in that last sentence are deliberate: They reflect the fact that scientists blush at the mention of ball lightning. They can't explain it properly and would rather it just went away. It's my job to explain natural phenomena, and I think it's just great that science can explain most things neatly, but not everything. I can't recall ever meeting anyone who had a strong claim to having seen ball lightning, and I certainly haven't yet. If you do, enjoy the moment, take some video if you'd like to, then dine out on it until you stop being invited to social events.

Lightning killed more than a hundred people in north and eastern India in the early monsoon season of 2020. Everyone knows it's dangerous, but few know what to do when caught closer to a storm than they planned to be. In an earlier book, *The Lost Art of Reading Nature's Signs*, I put a lot of time into honing practical safety advice about steps to take if you're worried about lightning. I'd never normally want to repeat a subject from one book to another, let alone repeat the words used, but for understandable reasons, I'll make an exception here, then go on to explore this area a bit further.

If you suspect lightning may be a risk, it is best to avoid open areas and find shelter: A car or shed will both be safer than standing in the open, providing you don't touch the metal parts. Move away from isolated tall objects, like solitary trees, and descend if safe to do so. Move away from water and get out of it if you're in it. Make sure you are not holding anything metal and if you deem the risk to be high, temporarily ditch anything that contains metal, like walking poles or a rucksack with a metal frame.

All I would add to that now is that if you're in the open, can't reach shelter, and are very worried, lower your body position but don't lie on the ground: A squat or crouch is best. You might find the 30/30 rule helpful: If the time between lightning and thunder is less than thirty seconds, head to a safe location. If you are indoors, wait thirty minutes from the last thunder before venturing out.

The exact spot that lightning chooses to strike is likely to be a prominent part of any landscape under the cumulonimbus, but we can't bank on this. The expression "a bolt from the blue," meaning something totally unexpected, has passed into everyday English, but it stems from lightning's ability to strike miles from the storm cloud, sometimes even appearing to come from a blue sky. This is rare, though. The risk is actually highest as the rain first reaches you.

Another popular idiom tells us that lightning doesn't strike twice. This may be true in love, war, or some other fields of human experience, but it's nonsense when it comes to weather. The Empire State Building was once struck fifteen times in fifteen minutes.

The color of lightning gives us a clue to the air it has passed through:

white = dry air

yellow = dusty

red = rain

blue = hail

Lightning generates electromagnetic waves that make it easier for forecasters to detect and track distant storms. These waves can be heard on AM radios as a snap or crackle.

STORM SYSTEMS

Storms can team up. Cumulonimbus clouds can interact and support each other. When they do, it can lead to marauding packs of storms, known in meteorology circles as mesoscale convective systems. If you experience storm after storm over several hours, one of these systems is passing through. Follow the guidance above, get indoors if you can, and do not attempt to read outdoor signs until you've had at least half an hour of calm weather.

TORNADOES

Tornadoes are unique whirling systems that range from strong winds to mayhem and destruction. These violent vortices start life as vigorous cumulonimbus clouds but need specific conditions that are much more likely at certain times of the year and in certain regions than others. The probability of tornadoes shoots up during May and June in the Great Plains.

The tornado's destructive energy is focused in a narrower band than in other storm systems, which has the twin effect of maximizing destructive power and reducing the area affected. This is why the path of these storms is so marked: Winds of 300 mph

(480 km/h) can destroy a whole street, while a few blocks away, there are still leaves on the trees.

Since tornadoes are triggered by storm clouds, the signs that point to especially powerful storm clouds are also signs that tornadoes are more likely. Prominent bulges at the top of the anvil or large mamma, both indicating unusually strong updrafts and downdrafts of air, are portents of trouble if you're in the wrong place at the wrong time.

The tuba or funnel cloud that forms at the bottom of the clouds is also a sign that conditions are ripe for a tornado. More often, even if these funnels stretch all the way down to the ground, a weaker relative of the tornado is formed: the landspout or waterspout, a whirling column of air and water.

HURRICANES

This type of storm is known variously as a hurricane, cyclone, typhoon, or tropical revolving storm, depending on the part of the world you're in. The names all refer to the same terrifying storm phenomenon, which here I will call a hurricane.

Like tornadoes, hurricanes are much more likely in specific regions and seasons. They always form in the tropics, and many regions don't see them at all, year round. Hurricanes are a hundred times larger than tornadoes and have a lot more total energy but are spread more thinly. Top wind speeds of hurricanes are lower than those of tornadoes, but they cause chaos over much wider areas.

Hurricanes need moisture, warm air, and sea to get going. The experience at ground level is violent and chaotic, but seen from the perspective of satellites, they are simple, elegant systems. Winds rotate, as they always do, counterclockwise around a low-pressure system in the northern hemisphere. Hurricanes are unusual only in the extremity of the low pressure and the way the system feeds itself, using water and the heat energy in the atmosphere to keep building in strength.

Most sailors crossing the Atlantic from Europe to the Caribbean sail in December, and this is no coincidence. The hurricane season in that band of the Atlantic runs from June to November, and the probability of being hit by one is reduced significantly in December. Before my first crossing of the Atlantic, I attended several weather briefings by experts in this sailing route. These days, all sailors wisely rely on others to forecast and warn of hurricanes, but the tradition is to look for certain signs, many of which we're already familiar with. Bad omens include sudden changes in wind strength and direction, the cloud progression that indicates storms, ocean swell from the direction of a suspected approaching storm, precipitous drops in air pressure, and a wave of heavy rain. These are the fundamentals that work all over the tropics. There are also intriguing local traditions and signs, such as "clouds like the skin of a cow," once used by sailors off the coast of India.

The most interesting nugget I gleaned and have never forgotten was that our experience of hurricanes depends greatly on which side of a hurricane hits us: They are not equal. If we are on the move, sailing or on land, we might find ourselves in a situation where we cannot escape the hurricane entirely, but we can choose which part hits us. Hurricanes have dangerous and less dangerous quarters—there are no perfectly safe areas. Allow me a strange analogy.

Imagine a person standing in an alleyway swinging a baseball at the end of six feet of rope. From above, the ball is rotating counterclockwise. If you're forced to pass too close to this person without ducking, you may get hit by the ball, but the experience is not the same on both sides. If you pass to the left of them, the ball is rotating toward you—it will hit you in the face very hard. But if you pass to their right, the ball will hit you more slowly, on the back of the head. Neither is pleasant, but the right side is less violent.

The winds in northern-hemisphere hurricanes are circling counterclockwise around the center of the low, like the ball above. If you are on a beach facing the Atlantic Ocean with a hurricane approaching, this means that the "danger quadrant"—in which the winds are fastest because they are rotating toward you—is to the left (north) of the center of the system.

Put another way, if you can't avoid a hurricane but you can move south so that the center of the hurricane heads a little to the north of you, your chances are better than if it heads just south of you.

If it headed straight over your head, there would be destruction from northerly winds, followed by an eerie calm as the "eye" went overhead, followed by more mayhem from southerly winds. The general advice is to avoid them if you can, and not just because working out where the "danger quadrant" is makes people's heads hurt.

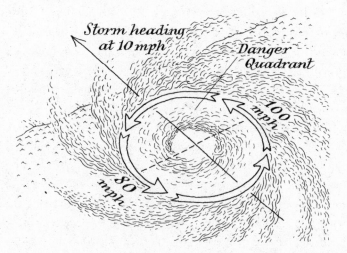

Hurricanes don't follow alleyways, but there are trends in their paths. They need the Coriolis effect, so they form at least three hundred miles (500 km) away from the equator. Once formed, they tend to head westward, averaging about 11 mph (18 km/h), and veer toward the nearer pole, which is why the Caribbean and the southeastern coast of the US are hit so regularly. And they sometimes curve back on themselves. If they head over cooler water or land, they lose the heat that powers them and their strength starts to fade.

When it comes to tornadoes, hurricanes, and other major havoc, we must accept our limitations. This is a time to tip our hats to the professional meteorologist and graciously acknowledge the value of their data, models, satellite photos, and experience.

CHAPTER 21

The Celestial and the Sublime

The Science and Art of Twinkling • Seasons and Stars • Halos •
The Grandest Shadows

THERE IS A DELIGHTFUL COLLECTION of traditional weather lore in a book known as *The Shepherd of Banbury's Rules*. It was written by "John Claridge, Shepherd" in the eighteenth century. It contains many familiar gems, including:

> When clouds appear like rocks and towers,
> The Earth's refresh'd by frequent showers.

There is meaning in all that we see, but still the temptation to invent meaning where there is none is too strong for some. It has been noted for millennia that the moon changes appearance markedly every few days and that the weather seems to change over a similar time frame. It proved just too tempting to draw a causal link between these two, so we find Mr. Claridge telling us, "A general Mist before the sun rises, near the full Moon—Fair Weather."

We know that an early-morning mist in summer is a good sign: Heat has escaped overnight because of clear skies. The moon has no role in this. When it comes to the moon's shape and phase, one of the shepherd's critics pointed out that there is a more appropriate saying:

The moon and the weather
May change together;
But change of the moon
Does not change the weather.
If we'd no moon at all,
And that may seem strange,
We still should have weather
That's subject to change.

The moon's phase can tell you about tides, time, and direction, but little or nothing useful about weather—at least that I know of. But that is not the same as saying the moon's appearance tells us nothing about the weather. The moon reflects the sun's light back toward us and shines a flashlight through the atmosphere. That journey can take the light through ice or vapor, and both change how we see the moon.

If we see an extraordinarily clear moon, it is a sign of dry, clear, and clean air. There are not only no clouds blocking the moon's light, but there is little moisture and therefore little prospect of clouds or rain in the coming hours. We know that clear night skies lead to heat loss through radiative cooling, which is why the old saying "Clear moon, frost soon" has a basis in science. The moon is not telling us anything directly about frosts but giving us a clue that the night will be much cooler than the day. Except for summer, frosts are more likely as a result.

THE SCIENCE AND ART OF TWINKLING

The farthest objects you will ever let your eyes focus on are the stars. They are not only the most distant things we will ever see, but they also offer one of the longest-range forecasts we can find in nature.

Light from the stars has to travel millions of miles to reach us, but the last tiny fraction of its journey has the greatest impact on how they appear. As the narrowest sliver of light hits our atmosphere it has to find its way through the air molecules, which are like syrup compared to the vast vacuum it has just crossed. The light is bounced and jostled a little by them, but if it bumps into any water molecules its journey is somewhat derailed. Stars that appear to twinkle a lot more than normal are telling us there is more water in the atmosphere. This can be one of the first signs of a front approaching, before even the cirrus appears.

It is about trends: Twinkling stars do not tell us much, but stars that twinkle noticeably more over the course of an evening or from one night to the next are a sign that moisture levels are increasing and that bad weather is probable. Stars always twinkle, or "scintillate," a bit, but relatively still stars suggest drier air, so fair weather and frosts are both more likely.

There is always an art within an art. The Pacific Island navigators used twinkling stars to forecast weather, and they distinguished between the way stars twinkled in different parts of the sky. They forecast that the weather changes would arrive from the part of the sky where the twinkling started.

An Australian meteorologist has claimed a link between meteors—shooting stars—and heavier-than-normal rainfall. It sounds far-fetched, but a simple logic lies behind the claim: Clouds form around tiny particles—nuclei—in the atmosphere, and the meteors shower dust into our atmosphere, which could provide abundant nuclei. He found a thirty-day delay from the major meteor showers, like the Geminids, to the spike in rainfall, which tallied with the time it would take for the dust particles to fall to the right level in the atmosphere.

Did you know that stars appear brighter in winter, even if the atmosphere is identical to a night in summer? We get to see a different part of the night sky, and therefore a different part of the universe, in each season. The stars we see in winter have a darker backdrop because there are fewer distant stars behind them. In winter we look out into emptier space but in summer toward the center of the Milky Way, our galaxy, which adds a faint background light to the night sky. Think how much easier it is to see flecks of white on a black, compared to grey, fabric. The effect is compounded because more of the brightest stars, like those in the luminous constellation Orion, are visible in winter than summer.

SEASONS AND STARS

In many indigenous cultures it is thought that the stars influence or control the weather. This is either true or false, depending on how we interpret these beliefs. If we take this concept to be literal, science does not support it. There is no known causal relationship between the stars we see and the weather we experience. But it is fair to assume that these societies are actually referring to the seasonal clock within the stars. For any cultures whose lives and livelihoods were tied to seasons, like the seafaring navigators of the Pacific, it was important to mark their passing in order to improve the chances of fair weather. As the British writer Arthur Grimble noted a century ago:

> Two constellations, the Scorpion and the Pleiades, punctuate the year for Gilbertese mariners. When the star Antares is in right ascension at sunset, then begins the fair-weather season, "the day of voyaging" (*te bongi ni borau*); it lasts until the Pleiades, appearing over the Eastern horizon with the first darkness, usher in the boisterous weather and close the travelling season. . . . These two periods correspond pretty accurately with the season of Trade winds and the season of westerly gales respectively.

We have seen how certain constellations indicate different seasons closer to home, like Orion, who hunts over the winter nights. Westerners wouldn't say that Orion brings snow, but we would all agree that snow is more likely in the months when Orion is visible. I think the reason for any divergent views lies in the difference between temperate and tropical climates and the resulting cultural perspectives.

The seasons are so markedly different in temperate zones that we sense seismic changes in temperature, the plants, and animals all around us, long before we might note any major star changes. In tropical climates, where temperature changes are smaller and the shift may be between fickle wet and dry periods, the relationship with the stars is likely to be more prominent. Tropical societies also see more of the night sky year round. We see little of it in midsummer or midwinter—it's too light in June and too cold in January for us to want to spend lots of time outside, neither of which is true in the tropics. In summary, while we feel the seasons change, often in our bones, tropical cultures wait for the stars to tell them. Neither believes that the stars are pulling the strings.

Some Arab cultures had a trickier time of it because the Muslim calendar is lunar, which does not align with the seasons, or the solar or star calendars. They had to devise their own relationship with time based on the number of days from the start of the year. Indian Ocean navigators would talk of sailing on *thalatha wa-tis 'in (fi) l-nairuz,* meaning the ninety-third day from the start of the year.

HALOS

A halo around the moon tells us that there is a thin ice cloud, most likely cirrostratus, and this can form part of a forecast. We know that cirrostratus after cirrus warns of a warm front approaching, so cirrus clouds followed by a halo around the moon warn us that rain is on its way. A halo around the sun means the same thing, cirrostratus, and forms part of the same forecast.

THE GRANDEST SHADOWS

At the end of a sunny day, look east just after sunset, and if you have a clear view, you may notice a darkening band rise very slowly from the horizon. This is Earth's shadow, and it is clearest when there is a little moisture or haze in the atmosphere. It can also be seen by looking west at dawn, but that's harder, as the atmosphere contains fewer particles.

The shadow appears as a veil of blue-purple color, rising at dusk or sinking at dawn. If you think you've spotted it, look for a pink band of light just above the shadow, known as the Belt of Venus, as this is where the planet is so often seen. It helps to have a low horizon, as near to sea level as possible, and to view it from a little height. Coastal slopes on the east side of sunny islands are made for it.

CHAPTER 22

Our Weather

I AM WRITING THIS FROM LOCKDOWN in May 2020. COVID-19 continues to scare and perplex, and yesterday more than fifteen hundred people died with this virus in the US alone.* The global media are running stories about sunlight having positive effects and hampering its spread. It is too early to try to understand the exact role the weather is playing in the life of this particular virus, but history suggests we can expect there to be a significant relationship.

In 1993, ten Navajo residents of the Four Corners region died with flu-like symptoms after their lungs filled with liquid. They were young and healthy before the illness struck. This was not the first time that the Navajo had suffered lethal outbreaks of this kind: There were similar reported cases in 1918 and 1933. It puzzled the medical community for many years.

---

* During the edit of this book, I was horrified to see how quickly this figure had come to seem small.

The oral historical tradition of the Navajo helped solve the mystery. The outbreaks all followed periods of unusually heavy rain or snow. It wasn't the rain or snow that was killing anyone, so at first the connection wasn't obvious. After speaking with the Navajo, investigators started to see how the pieces fit together. The precipitation had led to a massive increase in the number of pine nuts on the trees. This in turn led to an explosion in the number of rodents surviving. The Navajo victims were killed by Hantavirus Pulmonary Syndrome, following a spike in human–rodent contact. It was a sad tale of rain, nuts, rodents, virus, and death.

Thousands of miles away, in Siberia, Sooyong Park, a Korean photographer and naturalist, spent twenty years observing every move of the elusive Siberian tiger. Park noted how the wind carried the pine-nut pollen. The pollen would fertilize the flowers, and soon the pine nuts would grow and fall. The grazing animals sought out the nuts; the tigers followed their tracks and stalked their prey. The red pollen mapped the breeze and sketched a trail that Park could read. It led him to the tigers.

Thousands of miles from Siberia and the Navajo, in Denmark, scientists figured out that they could use the air temperature to predict how many prey tiger beetles would catch.

From rain and wind through pine nuts, tigers, and tiger beetles, it is clear that weather, climate, and microclimate are integral parts of the jigsaw puzzle of nature. The closer someone is to the land, the more they appreciate its relationship with the sky. Russian peasants traditionally looked for links between the blossoming time of cherry trees and frosts and between the color of a bird's plumage and the rain heading their way.

The Scottish writer Nan Shepherd put it well: "The disintegrating rock, the nurturing rain, the quickening sun, the seed, the root, the bird—all are one."

From her small farm in the Hebrides, the writer Tamsin Cal-idas sensed that the weather would change. The sheep sheltered in the woods nearby and the gulls huddled close to the feeders. But she also felt it within her: "The dull leaden tang in my mouth. . . . That blunt-edged taste is a harbinger of either trouble or change."

Weather forms a vital part of our inner world. Shorter winter days trigger Seasonal Affective Disorder in millions, and, unsurprisingly, it is seven times more prevalent in Alaska than it is in Florida. The sudden cold breeze gives us a physical chill but also a sense that it is time to head for home. And when we do, our feet march to the drum of the wind. Researchers found that people in a town walked at normal speeds until the wind reached force 6, when there was a sudden acceleration in those studied. But we don't all react identically. The wind is sexist.

Women's breasts are more vulnerable to cold than men's chests, and a blast of cold wind can stop the flow of milk in a nursing mother. This may explain why men tend to face into strong winds, whereas women tend to face away. The study behind that finding was published in 1976 and, although the same winds still blow, a lot has changed since then about how we feel about ourselves. If the study were repeated, I wonder if we could uncover how much of this difference in behavior is biological and how much is cultural. Either way, mothers with babies will doubtless still be the first to find the warm pockets in spring and guide you toward the best shelter, long before anyone else thinks to look.

When the cold wind hits us, we respond automatically, even if we're bundled up and don't feel colder. Studies have shown that our heart rate drops if cold air blows onto our forehead for only thirty seconds, as the body prepares to conserve heat nearer the core.

Weather, whether it is kind or causing trouble, shapes who we are and has done so from the beginning. Some historians have argued that the great early civilizations—the Phoenicians, Egyptians, Assyrians, Babylonians, Chinese, Aztecs, Maya, and Inca—all sprouted in places where the average temperature was close to what we now consider comfortable indoors, 68°F (20°C).

Weather and climates build empires over eons, but micro-climates create the wondrous moment. Why are there so many vineyards in this cool valley? Ah, the sun is bouncing off the river and giving the vines a double dose of its light. Who could not feel a wave of joy in such a discovery? How many delights are hidden under our noses? Will we ever notice that rainbows are a tiny bit smaller at the coast because of the salt in the rain? The answer is not as important as the question. The act of looking brings wonder.

Sources

p. 4 "FitzRoy grew depressed, and took his life in 1865": W. Burroughs et al., p. 73.

p. 4 "Very little degree of accuracy can be guaranteed for any forecast . . .": R. Lester, p. 123.

p. 7 "The climate on two sides of a 2,600-foot-high (800 m) ridge in the Swiss Jura mountains is so different": Ph. Stoutjesdijk and J. J. Barkman, pp. 77–78.

p. 8 "The difference in climate between the north and south sides of juniper bushes in temperate zones of the US and Europe": Ph. Stoutjesdijk and J. J. Barkman, p. 7.

p. 8 "Heathland loses heat very quickly at night and can easily be around 5°F (3°C) colder": Ph. Stoutjesdijk and J. J. Barkman, p. 55.

p. 18 "they preferred to attempt to fly near the start of the day": E. Sloane, p. 88.

p. 18 "small spiders which swarm in the fields in fine weather in autumn, and have a power of shooting out webs . . .": G. White, p. 176.

p. 18 "And in the 1830s, Charles Darwin noticed spiders reaching his ship, the *Beagle*, even though it was one hundred yards off Argentina": irishtimes.com/news/environment/how-ballooning-spiders-fly-through-the-sky-1.4000228 (Accessed 11/28/19).

p. 19 "there is some evidence that if the heating of the land is too intense and the thermals too powerful": M. H. Greenstone.

p. 19 "Animal-behavior experts have known for at least a century that bird weight, thermals, and time of day are all related": R. S. Scorer.

p. 27 "*kapesani lang*": S. Thomas, p. 297.

p. 28 "Dark patches of cloud—spots of ink on the sky, the 'messengers'—go drifting by . . .": R. Jefferies, pp. 34–35.

p. 29 "When the clouds are upon the hills . . .": J. Claridge, p. 47.

p. 33 "The rounded bulges at the top are a sign that the air is still rising": S. Dunlop, p. 70.

p. 35 "Clouds are lower over oceans than over land": U. Lohmann et al., p. 4.

p. 36 "There are rarely any cumulus clouds in Antarctica": W. Burroughs et al., p. 194.

p. 37 "Thermal detection is a Zen-like art that brings together all your senses . . .": xcmag.com/news/zen-and-the-art-of-circles-part-1 (Accessed 1/14/20).

p. 44 "Ninety percent of the water vapor in the air comes from the oceans": W. Burroughs et al., p. 39.

p. 44 "In October 2019 Denver, Colorado": ITV, 10/19/19.

p. 44 "The Wola people of Papua New Guinea . . . *Chay nat*": "Wola People, Land and Environment" in P. Sillitoe.

p. 46 "Those off Peru can affect Australia, a phenomenon known as El Niño": W. Burroughs et al., p. 38.

p. 46 "many places in the world at the same latitude have totally different climates, like Edinburgh and Moscow": R. Lester, p. 130.

p. 59 "As the wind grows from nothing to the very gentlest breeze . . . Swifts will stay airborne even as branches break": L. Watson, p. 217.

p. 62 "In drought-sensitive parts of the world, like the southwestern US": R. Inwards, p. 74.

pp. 62–63 "George Washington kept a detailed weather diary, and it may not be stretching things": M. Lynch, p. 31.

p. 63 "Every wind has its weather.": R. Inwards, p. 68.

p. 63 "Out of the south cometh the whirlwind.": R. Lester, p. 141.

p. 64 "A veering wind, fair weather . . .": L. Watson, p. 78.

p. 65 "If the wind shifts from south to north through west, there will be . . .": R. Inwards, p. 74.

p. 78 "Judges 6:36–8": from Berean Study Bible, biblehub .com/judges/6-37.htm (Accessed 2/24/10). First seen in Ph. Stoutjesdijk and J. J. Barkman, pp. 59–62.

p. 80 "One August morning in 1986, police in Pennsylvania received a call from a distressed man": peoplemagazine.co.za/real-people/real-stories/meet-mr-duplicity and youtube.com/watch?v=8rkfev2OXls (Accessed 3/2/20).

p. 82 "The leaves, grass and twigs became encased in cylinders of ice, so that the trees swayed in the slightest breeze . . .": W. P. Hodgkinson, p. 92.

p. 84 "the inversion layer during a sharp frost is so robust that commercial fruit growers sometimes use it like a greenhouse roof": goodfruit.com/the-frost-fight (Accessed 2/26/20).

p. 85 "Dr Hales saith, 'That the warmth of the earth, at some depth under ground, has an influence in promoting a thaw . . .'": G. White, p. 20.

p. 86 "Grasslands will be much colder and frostier than fields, but heaths, dry reed beds, and drained peat bogs": J. P. M. Woudenberg.

p. 88 "it is stopped in its journey by a railway embankment on the other side of the hollow": weatheronline.co.uk/reports/ wxfacts/Frost-hollow.htm (Accessed 2/26/20).

p. 89 "Scientific experiments have shown that a polystyrene box only 3 feet by 3 feet (1 m by 1 m) creates a shockingly different microclimate": Ph. Stoutjesdijk and J. J. Barkman, p. 58.

p. 89 "The indigenous Wola people of Papua New Guinea . . . rejoicing in frosty schadenfreude has its own name, *liywakay*": P. Sillitoe, p. 74.

p. 91 "Theophrastus, the ancient Greek philosopher, noted how coastal rain in Greece tasted salty": Theophrastus, Sign 25, p. 21.

p. 91 "Sadly, a study in the Rocky Mountains in Colorado found that 90 percent of rainwater samples": earthsky.org/earth/rain- microplastic-rocky-mountains-colorado (Accessed 5/3/20).

p. 98 "once there are more heaped clouds than blue sky, showers are likely": S. Dunlop, p. 111.

p. 99 "In some parts of the Scottish Highlands, the windward west sides receive six times as much rain as the eastern lee sides": R. Lester, p. 38.

p. 103 "Many mountain ranges have rain shadows on their eastern side": W. Burroughs et al., p. 37.

p. 102 "foehn gap": K. Stewart, p. 264.

p. 104 Miami vs Seattle climate: en.wikipedia.org/wiki/
List_of_cities_by_sunshine_duration, weather.com/
science/weather-explainers/news/seattle-rainy-reputation,
en.wikipedia.org/wiki/Climate_of_Miami (All accessed 3/9/20).

p. 105 Raindrop size and formula from "Meso-Micro," Lohmann et al,
p. 209.

p. 122 "onion layer": R. Lester, p. 41.

p. 127 "When the icy wind warms, expect snow storms.": O. Perkins, p.
92.

p. 127 "You may notice that bridges and overpasses allow snow to
accumulate before the lower roads nearby": theweatherprediction.
com/habyhints/201 (Accessed 3/10/20).

p. 130 "those moist pockets are havens for spring flowers": Ph.
Stoutjesdijk and J. J. Barkman, p. 67.

p. 130 "Trees in snowy and exposed climates, like the pines of Lapland":
J. J. Barkman, 1951.

p. 131 "In many regions organisms paint tree bark and rocks with clues":
R. Nordhagen.

p. 131 "Corn is as comfortable under the snow as an old man . . .": R.
Inwards, p. 115.

p. 133 Białowieza Forest: Ph. Stoutjesdijk and J. J. Barkman, p. 72.

p. 131 "If you find ice over a pond or lake that is thick enough to walk
across, beware": Ph. Stoutjesdijk and J. J. Barkman, p. 70.

p. 131 "Dwarf willow, for example, is a sign that we can expect at least
two months without snow cover each year": Ph. Stoutjesdijk and J. J.
Barkman, p. 71.

p. 132 The Calhoun, Tennessee, accident: all sourced from W. Haggard,
pp. 99–116.

p. 136 "Whenever there is fog, there is little or no rain": W. Burroughs et
al., p. 65. And Theophrastus.

p. 137 "A summer fog is for fair weather . . . will freeze a dog": R.
Inwards, pp. 8–9.

p. 138-139 "Forests act as fog-catchers and have been planted in places
like Japan, where sea fogs cause problems in coastal areas": J. Grace,
p. 179.

p. 139 "but if there is no wind, fogs can linger in the cool woodland for
longer than they do outside": R. Lester, p. 59.

p. 141 "But if visibility remains poor after rain, there's more to come": O. Perkins, p. 69.

p. 141 "call boys": S. Dunlop, p. 69.

p. 150–51 "And it snakes on its journey; it may flow from south-southeast to north-northwest, or south-southwest to north-northeast, but it tends towards a west-to-east path": Chris Mcconnell and Peter Gibbs, personal correspondence.

p. 153 "This means that an increase in winds is likely in the next twelve hours": O. Perkins, p. 29.

p. 153 "The sign is especially strong if the alignment of the ropes is from northwest to southeast": S. Dunlop, p. 94.

p. 156 "Contrails that not only survive for long periods but appear to grow, widening and spreading, are a sure sign of an atmosphere near saturation and a strong clue that wet weather is approaching": M. Kästner, R. Meyer and P. Wendling.

p. 157 "The vortices pull the contrails into two separate lines, which merge soon after": R. Stull, p. 166.

p. 158 "Fallstreak Holes . . . Sometimes when a hole is punched through the cloud, the broken pieces come back together and form a new cloud": S. Dunlop, p. 94.

p. 160 "The space between cloud streets is usually between two and three times their height": weatheronline.co.uk/reports/wxfacts/Cloud-streets.htm (Accessed 4/1/20).

p. 166 "more likely to be nearer 160 feet (50 m) in the Arctic": U. Lohmann et al., pp. 8–9.

p. 186 "For a wind passing through a gap is always more forceful and vigorous like a current . . .": J. Morton, p. 61.

p. 187 "It was like being under attack. I crossed the street in defensive posture, shielding my eyes with a chilled hand . . .": N. Hunt, p. 200.

p. 187 "The Santa Ana winds in California vary from light breezes near their birth at the high-pressure end to more than 60 mph (97 km/h) when they are pinched between the coastal mountains of the San Gabriel and San Bernardino ranges": *Meteorology for Naval Aviators*, p. 21.

p. 189 "Opposite Helike the Bear there is a foreland called Karambis, steep on every side . . .": J. Morton, p. 57.

p. 192 "There is not a lot to be gained by knowing that the eddy pattern downwind of Douglas firs is different from that of spruces": J. Grace, p. 23.

pp. 193–94 "Rotor streaming": J. Grace, p. 151.

p. 194 "In 1986 a gust of 173 mph (278 km/h) was recorded at the summit of Cairn Gorm in the Scottish Highlands": P. Eden, p. 52.

p. 195 "The summits with the strongest relative winds are the tallest, narrowest, most isolated peaks nearest the coast": J. Grace, p. 149.

p. 195 "from half to more than three times those of the main wind nearby": J. Grace, p. 152.

p. 196 "The Palmdale Mountain Wave knocks downs trees and rattles windows in Palmdale, California": losangeles.cbslocal.com/2019/01/07/palmdale-mountain-wave (Accessed 4/7/20).

p. 196 "In November 1950, an elderly patient named Edward Nevin was recuperating at his San Francisco home": whole story, W. Haggard, pp. 11–17.

p. 198 "It only kicks in once the land has warmed at least 9°F (5°C) more than the sea": R. Stull, p. 654.

p. 200 "Where sea breezes meet over a peninsula, the two fronts can join to form a line of clouds": O. Perkins, p. 79.

p. 200 "In summer, a fog that appears along the coast and doesn't seem to fit the weather": A. Watts, p. 34.

pp. 201–2 "The dark, steep slopes of the volcanic hills in the Atacama Desert in Chile warm quickly in the morning sun": Cloud Appreciation Society newsletter, April 2020.

p. 202 "a cold mountain breeze may still be flowing down from the summit": J. Grace, p. 152.

p. 202 "Lake Garda": L. Watson, p. 36.

p. 203 "Theophrastus and even the mighty Aristotle": J. Morton, p. 55.

p. 203 "The rolling swell of the sea had been aroused by the strong breezes which blow . . .": J. Morton, p. 55.

p. 204 Molière, *Le Bourgeois Gentilhomme*: en.wikipedia.org/wiki/Tramontane (Accessed 11/10/20).

p. 204 "*golobica* . . . You can map the route of the Bora by following the stone doves": N. Hunt, p. 75.

p. 209 "The temperature drop has been measured to as much as 7°F (4°C)": A. J. Van der Poel and Ph. Stoutjesdijk.

p. 280 Microclimate under trees: Full list of many sources can be found in Ph. Stoutjesdijk and J. J. Barkman, pp. 96–99.

p. 210 "Holm oaks will tolerate very cold weather, but not for long periods. They can survive a burst of −4°F (−20°C), but not long periods of even 30°F (−1°C)": Barkman, cited in Ph. Stoutjesdijk and J. J. Barkman, p. 6.

p. 213 "The Wind Bulge": J. Grace, p. 19.

pp. 217–18 Slocum quote: gutenberg.org/files/6317/6317.txt (Accessed 4/15/20).

p. 219 "Trees are surprisingly sensitive to wind; if they are shaken for only thirty seconds each day they grow to be 20 to 30 percent shorter than a sheltered tree nearby": P. L. Neel and R. W. Harris.

p. 221 "Thistle seeds only need the lightest of breezes, 2 mph (3 km/h), to sail through the sky . . . that it can take them 8 seconds to fall 3 feet (1 m)": L. Watson, pp. 172–73.

p. 222 "they are often frost pockets . . . woodland gaps can receive 50 percent more snow than a nearby open field": R. Geiger.

p. 223 "To dwellers in a wood almost every species of tree has its voice as well as its feature . . .": bbc.co.uk/programmes/m000b6sm (Accessed 9/3/20).

p. 223 "Another author wrote of an apple tree being a cello, an old oak a bass viol, a young pine a muted violin": Guy Murchie quoted in L. Watson, p. 263.

p. 223 "Among plants and trees, those with large leaves have a muffled sound; those with dry leaves have a sorrowful sound . . .": awatrees. com/2013/01/06/psithurism-the-sound-of-wind-whispering-through-the-trees (Accessed 6/17/20).

p. 227 "In an experiment dating back over a century, fruit growers placed a cut-out letter on a ripening apple": S. Elliott, p. 39.

p. 229 "Through certain plants, I can predict whether it will rain or not": O. D. Kolawole et al., "Ethno-Meteorology and Scientific Weather Forecasting: Small Farmers and Scientists' Perspectives on Climate Variability in the Okavango Delta, Botswana," *Climate Risk Management* 4–5 (2014): pp. 43–58.

p. 229 "The Wola in Papua New Guinea study the behavior of a common sword grass they call *gaimb*": P. Sillitoe.

p. 229 "The peasant farmers of Tlaxcala, Mexico, keep an eye on izote, the yucca plant": A. D. Rivero-Romero et al., "Traditional Climate Knowledge: a Case Study in a Peasant Community of Tlaxcala, Mexico," *Journal of Ethnobiology and Ethnomedicine* 12 (2016).

p. 231 "It snowed in Miami in January 1977": Google search for "snow in Miami" (Accessed 4/20/20).

p. 232 "forests are normally found on the northern slopes of the hot, dry Atlas mountains": Ph. Stoutjesdijk and J. J. Barkman: p. 81.

p. 232 "their catkins flower on their sunny south side first": G. Kraus.

p. 232 "Edith's checkerspot in North America": S. B. Weiss, D. D. Murphy and R. R. White.

p. 234 "dandelions, bindweed, buttercups, tulips, crocus, daisies, marigolds, Carline thistles, gentians, red sand spurrey, and wild indigo": R. Inwards, pp.154–55 and elsewhere.

p. 234 "Tulips and crocus close when the temperature drops . . . members of the *Silene* family": W. G. van Doorn and U. van Meeteren, "Flower Opening and Closure: A Review," *Journal of Experimental Botany* 54 (2003): pp. 1801–12.

p. 235 "The common cock's-foot grass moves very slowly toward the wind": M. L. Luff.

p. 235 "Matgrass (*Nardus stricta*) . . . this creates a 'horseshoe' shape in the grass: The two thin ends are aligned to point roughly north and the curved part is at the southern end": Ph. Stoutjesdijk and J. J. Barkman, p. 79.

p. 236 "Bracken has been found in fossils dating back fifty-five million years": B. Myers, p. 35.

p. 238 "*Cornus suecica*": Ph. Stoutjesdijk and J. J. Barkman, p. 3.

p. 239 "Watch hawthorns mark spring. The branches nearest the ground come into flower well before the higher ones": Ph. Stoutjesdijk and J. J. Barkman, p. 160.

p. 239 "snowdrops, bluebells, fig buttercup, and muskroot . . . burst into life first in the places that warm quickly in early spring": R. Geiger.

p. 240 "The flower stalks grow shorter and the number of flowers decreases as you climb any mountain": J. Grace, p. 143.

p. 240 "The farmed grass is cool enough to keep grasshopper eggs from hatching": W. K. R. E. van Wingerden and R. van Kreveld.

p. 241 "One researcher discovered that the leaves of wolfsbane (*Arnica montana*) halved in size when he climbed from 6,000 to 8,000 feet (1,900 to 2,450 m)": Werger, see Ph. Stoutjesdijk and J. J. Barkman, p. 164.

p. 242 "Even the curvature of leaves offers clues to the microclimate": M. J. A. Werger, pp. 123–60.

p. 243 Fungi timing: S. Pinna, M-F. Gevry and M. Cote.

p. 244 "Some scientists believe that many fungi species eject spores in response to a drop in air pressure": newscientist .com/article/dn19503-fungi-generate-their-own-mini-wind-to-go-the-distance (Accessed 4/24/20).

p. 245 "Researchers in the Netherlands made an amazing discovery. They figured out that you can tell how many foggy days": Ph. Stoutjesdijk and J. J. Barkman, p. 5.

p. 246 "Black-spotted sycamore leaves are down, but the moss grows thick and deeply green . . .": R. Jefferies, p. 183.

p. 246 Blackberries: A. Young et al.

p. 247 "Southwestern Ireland is mild enough for palms to grow, so we might assume that wheat would thrive there, but it doesn't: It's too cool and wet in summer": Ph. Stoutjesdijk and J. J. Barkman, p. 6.

p. 247 "*honami*": E. Inoue.

pp. 261–62 Monroe and Venturi Effects: L. Watson, p. 228.

p. 263 "The exhausts from cars are blown towards the leeward side of the street, while the downwind side gets fresher air from above": *Air Quality in Cities*, N. Moussiopoulos, ed., Springer: 2003.

p. 263 "I can see a couple of buzzards from my window . . .": Ben Davis, personal conversation on 9/21/20.

p. 264 "can make cities 22°F (12°C) warmer than the surrounding countryside": R. Stull, p. 678.

p. 264 "If a light main wind is blowing, the downwind side of a city will be a couple of degrees warmer than the windward side": R. Stull, p. 678.

p. 266 Walkie-Talkie building: bbc.co.uk/news/ uk-england-london-23930675 (Accessed 7/16/20).

p. 266 "hexagonal or octagonal buildings stand a better chance against brutal gales. And roofs with several angles are more resistant than a simple up-and-down gable roof": sciencedaily .com/releases/2008/07/080709110842.htm. (Accessed 7/17/20).

p. 266 "In parts of the Middle East, wind towers reach above the street to catch and channel any breezes": W. Burroughs et al., p. 133.

p. 266 "Hyderabad": L. Watson, p. 226.

p. 269 "'When stones sweat, rain you'll get.' . . . home runs": M. Lynch, p. 142.

p. 269 urban lore, e.g.: "mats swell, cheese softens . . .": R. Inwards, pp. 158–59.

p. 270 "The Six Winds of the City Brave": coursera.org/ lecture/sports-building-aerodynamics/ 4-1-wind-flow-around-buildings-part-1-PvfFX (Accessed 3/28/19) and P. Moonen et al., "Urban Physics: Effect of the Micro-Climate on Comfort, Health and Energy Demand," *Frontiers of Architectural Research* 1 (2012): pp. 197–228.

p. 273 "coral that exudes a clear liquid in advance of settled weather and a milky one before storms": D. Lewis, *Voyaging*, p. 125.

p. 274 *"Kia kohu te mata o Havaiki!* ('May the peaks of Havaiki be banked in clouds!')": D. Lewis, *We, the Navigators*, p. 221.

p. 275 "the *awwal al-kaws* and the *akhir al-kaws*": A. Constable and W. Facey, p. 76.

p. 276 "Continual winds from the north-west, which blow nearly all the year round . . .": A. Constable and W. Facey, p. 76.

p. 278 *"te nangkoto"*: D. Lewis, *We, the Navigators*, p. 216.

p. 283 "Never has rain brought ill to men unwarned. Either, as it gathers, the skyey cranes . . .": Virgil, *Georgics*, loebclassics .com/view/virgil-georgics/1916/pb_LCL063.125.xml (Accessed 5/7/20).

p. 284 "Male adders": Ph. Stoutjesdijk and J. J. Barkman, p. 156.

p. 284 "Another theory is that cows lie down more in the afternoon, which is when most showers fall": Simon Lee, personal correspondence.

p. 285 "Spiders' webs": F. Vollrath, M. Downes and S. Krackow, and jeb. biologists.org/content/216/17/3342 (Accessed 4/10/20).

p. 286 "Gabriel Oak": Hardy, *Far from the Madding Crowd*: etheses .whiterose.ac.uk/14104/1/479514.pdf (Accessed 5/5/20).

p. 286 "A horse-whisperer I know told me that the wind changes a horse's behaviour": Adam Shereston, personal conversations.

p. 287 "Slugs": jeb.biologists.org/content/jexbio/31/2/165.full .pdf (Accessed 5/5/20).

p. 288 "From January, if nighttime temperatures rise above 41°F (5°C), frogs and toads emerge from hibernation": froglife.org/info-advice/ frequently-asked-questions/ frogs-and-toads-behaviour (Accessed 11/20/20).

p. 288 "Frogs grow more active with rising humidity": E. D. Bellis. "The Influence of Humidity on Wood Frog Activity," *The American Midland Naturalist* 68, no. 1 (July 1962): pp. 139–48.

p. 288 Earthworms: scientificamerican.com/article/ why-earthworms-surface-after-rain (Accessed 6/5/20).

p. 289 "Through the chirpings of some birds and sounds from certain insects, I can predict . . .": O. D. Kolawole, *Climate Risk Management*, pp. 43–58.

p. 289 Red kites: Simon Lee, personal correspondence.

p. 289 "On hot, calm days, birds sound farther away than they are": C. S. Robbins, "Bird Activity Levels Related to Weather," *Studies in Avian Biology* 6 (1981): pp. 301–10.

p. 290 "birds gliding, but they struggle to do this in turbulent air": Ph. Stoutjesdijk and J. J. Barkman, p. 27.

p. 290 "In Pennsylvania, the prevailing wind hits the mountains": E. Sloane, p. 69.

p. 290 "The more I think, the more I am convinced that the buoyancy of air . . .": R. Jefferies, p. 123.

p. 291 "Birds sunbathe to warm up, and if they take off from a sunny spot, they're likely to be heading for another sunny spot . . . a bird like a jackdaw with its beak open, panting": Ph. Stoutjesdijk and J. J. Barkman, pp. 147–51, 200.

p. 292 "'Seem' is important: . . . Are the owls really making more noise or has the calm weather quietened the trees, making them seem louder?": Simon Lee, personal correspondence.

p. 292 "Francis Bacon": R. Inwards, p.138.

p. 292 Bats and insects: Rupert Lancaster, personal correspondence and K. Parsons, G. Jones, and F. Greenaway.

p. 292 "One theory is that when prevailing winds bend the trees over toward the northeast": woodpecker-network.org .uk/index.php/news/51-great-spotted-woodpecker-nest-hole-orientation (Accessed 5/7/20).

p. 293 Bumblebees, "Honeybees do not leave the hive to forage for food until it is 55°F (13°C) outside, and they grow more active": en.wikipedia.org/wiki/Forage_(honey_bee) (Accessed 5/6/20) and Z. Puškadija et al., "Influence of Weather Conditions on Honey Bee Visits (*Apis Mellifera Carnica*) During Sunflower (*Helianthus Annuus L.*) Blooming Period," *Agriculture: Scientific and Professional Review* 13, no. 1 (2007): p. 13.

p. 293 Bumblebees: D. M. Unwin and S. Corbet, p. 20.

p. 293 Beetles: H. Dreisig.

p. 293 "Big insects prefer the cool periods earlier and later in the day and will be less active in the middle": D. M. Unwin and S. Corbet, p. 24.

p. 294 "If you see a swarm, they're telling you that it's more than 55°F and the winds are slower than 20 feet (6 m) per second": rsb.org.uk/ get-involved/biology-for-all/flying-ant-survey (Accessed 9/7/20).

p. 294 "scientific research supports the idea that many ants are sensitive to humidity—weaver ants have been observed building nests before tropical storms": S. Bagchi, "Weaver Ants as Bioindicator for Rainfall: An Observation," researchgate.net/publication/277126182_ Weaver_ants_as_bioindicator_for_rainfall_An_observation (Accessed 10/11/20).

p. 294 "more than thirteen thousand named ant species on the planet": cosmosmagazine.com/biology/can-ants-predict-rain-yes-no-maybe (Accessed 12/5/20).

p. 295 "Morning is a good time for studying butterflies": J. Lewis-Stempel, p. 145.

pp. 295–96 Butterflies and temperature: L. Wikström, P. Milberg and K.-O. Bergman.

p. 296 "Butterflies are also sensitive to the sun's radiation and are more likely to start flying when they are in direct sunlight . . . Butterflies are less active in cloudy conditions, flying over shorter distances and times": A. Cormont et al., "Effect of Local Weather on Butterfly Flight Behaviour, Movement, and Colonization: Significance for Dispersal Under Climate Change," *Biodiversity and Conservation* 20 (2011): pp. 483–503.

p. 296 "battling over sun flecks in shady woods": D. M. Unwin and S. Corbet, p. 18.

p. 297 "Next morning the chinook wind was gone and with it the flies": M. Schaffer, p. 83.

p. 297 "Amazingly, there is some evidence that on windy days, biting insects, including mosquitoes, bite prey on the lee side": Ph. Stoutjesdijk and J. J. Barkman, p. 66.

p. 298 "There are about 150 species in the UK and more than 700 across the US and Canada, and most are less than 2 millimeters long": M. Chinery, p. 20, and R. Foottit, "Thysanoptera of Canada," *ZooKeys* 819 (January 2019): 289–92.

p. 298 Thrips and electricity: thrips-id.com/en/thrips/thunder-flies (Accessed 11/22/20).

p. 298 "If not wheeling in the sky, look for them over the water, the river, or great ponds . . .": R. Jefferies, pp. 30–31.

p. 300 "At any one time, about two thousand thunderstorms are raging around the world": R. Stull, p. 564.

p. 305 "In 1975 a particularly impressive local storm dumped three months' worth of rain on a patch of north London in a few hours": P. Eden, p. 131.

p. 306 "two fine days and a thunderstorm": P. Eden, p. 107.

p. 307 "A thunderstorm in April is the end of the hoar frost.": R. Inwards, p. 24.

p. 308 "They can be travelling at 30 mph (48 km/h) vertically upwards": W. Burroughs et al., p. 199.

p. 310 "In 1938 five glider pilots were searching for rising air during a competition": L. Watson, p. 49.

p. 310 "Flight 191": W. Haggard, p. 41 onwards. And en.wikipedia.org/wiki/Delta_Air_Lines_Flight_191 (Accessed 5/12/20).

p. 313 "The darker regions are the updrafts and the lighter, rougher regions are where downdrafts and heavy rain are": R. Stull, p. 482.

p. 313 "Lightning heats the air to approximately 54,000°F (30,000°C)": P. Eden, p. 7.

p. 315 "Lightning killed more than a hundred people in north and eastern India": theguardian.com/world/2020/jun/26/lightning-strikes-kill-more-than-100-in-india (Accessed 7/20/20).

p. 316 "If you suspect lightning may be a risk, it is best to avoid open areas . . .": T. Gooley, *The Lost Art of Reading Nature's Signs*, p. 150.

p. 316 "30/30 rule": R. Stull, p. 570.

p. 316 "a bolt from the blue": S. Dunlop, p. 114.

p. 316 "The risk is actually highest as the rain first reaches you": R. Stull, p. 567.

p. 316 "The Empire State Building was once struck fifteen times in fifteen minutes": W. Burroughs et al., p. 241.

p. 316 "The color of lightning": W. Burroughs et al., p. 240.

p. 317 "Lightning generates electromagnetic waves that . . . can be heard on AM radios as a snap or crackle": R. Stull, p. 568.

p. 317 "Tornadoes": W. Burroughs et al., p. 244.

p. 319 "clouds like the skin of a cow": G. R. Tibbetts, p. 385.

p. 319 Traditional hurricane signs: G. R. Tibbetts, p. 385.

p. 321 "they tend to head westward, averaging about 11 mph (18 km/h), and veer toward the nearer pole": S. Dunlop, p. 126.

p. 322 "When clouds appear like rocks and towers . . .": J. Claridge, p. 35.

p. 322 "A general Mist before the sun rises, near the full Moon—Fair Weather.": J. Claridge, p. 48.

p. 323 "The moon and the weather, May change together . . .": J. Claridge, p. 50.

p. 323 "Clear moon, frost soon": J. Claridge, p. 12, W. Burroughs et al., p. 70 and passim.

p. 324 "The Pacific Island navigators used twinkling stars to forecast weather, and that they distinguished between the way stars twinkled in different parts of the sky": D. Lewis, *Voyaging*, p. 125.

p. 324 "An Australian meteorologist has claimed a link between meteors—shooting stars—and heavier-than-normal rainfall": E. G. Bowen.

p. 325 "The stars we see in winter have a darker backdrop because there are fewer distant stars behind them": earthsky .org/astronomy-essentials/star-seasonal-appearance-brightness (Accessed 5/19/20).

p. 326 "Two constellations, the Scorpion and the Pleiades, punctuate the year for Gilbertese mariners . . .": Arthur Grimble, from "Canoes in the Gilbert Islands," *The Journal of the Royal Anthropological Institute* 54 (1924): pp. 101–39.

p. 327 *"thalatha wa-tis 'in (fi) l-nairuz"*: G. R. Tibbetts, p. 361.

p. 330 "The oral historical tradition of the Navajo helped solve the mystery": daily.jstor.org/solving-a-medical-mystery-with-oral-traditions (Accessed 5/20/20).

p. 330 "Sooyong Park": S. Park, p. 24.

p. 330 "Russian peasants traditionally looked for links": V. Rudnev, "Ethno-Meteorology: A Modern View about Folk Signs," horizon .documentation.ird.fr/exl-doc/pleins_textes/divers18-08/010029408.pdf.

p. 330 "The disintegrating rock, the nurturing rain, the quickening sun . . .": N. Shepherd, p. 48.

p. 331 "The dull leaden tang in my mouth . . .": Tamsin Calidas, quoted *Sunday Times, Culture*, 5/17/20, p. 22.

p. 331 "seven times more prevalent in Alaska than it is in Florida": S. Nolen-Hoeksema, *Abnormal Psychology* (6th ed.), New York: McGraw-Hill Education, 2014, p. 179, from en.wikipedia.org/wiki/Seasonal_affective_disorder (Accessed 5/20/20).

p. 331 "Researchers found that people in a town walked at normal speeds until the wind reached force 6": S. L. Carson.

p. 331 "Women's breasts are more vulnerable to cold than men's chests . . . women tend to face away": B. Palmer.

p. 331 "mothers with babies will doubtless still be the first to find the warm pockets in spring": R. Geiger.

p. 331 "Studies have shown that our heart rate drops if cold air blows onto our forehead for only thirty seconds": J. Le Blanc.

p. 332 "Some historians have argued that the great early civilizations—the Phoenicians, Egyptians, Assyrians, Babylonians, Chinese, Aztecs, Maya, and Inca": S. F. Markham.

p. 332 "the sun is bouncing off the river and giving the vines a double dose of its light": O. H. Volk.

p. 332 "rainbows are a tiny bit smaller at the coast": R. Stull, p. 833.

Selected Bibliography

Barkman, J. J. "Impressions of the North Swedish Forest Excursion." *Vegetatio*, 3 (1951): pp. 175–182.

———. "The Investigation of Vegetation Texture and Structure." From M. J. A. Werger, *The Study of Vegetation*, Junk, The Hague, 1979.

Barry, Roger and Peter Blanken. *Microclimate and Local Climate*. Cambridge University Press, 2016.

Binney, Ruth. *Wise Words and Country Ways*. David & Charles, 2010.

Bowen, E. G. "Lunar and Planetary Tails in the Solar Wind." *Journal of Geophysical Research* 69, no. 23 (1964): pp. 4969–70.

Burroughs, William et al. *Weather: The Ultimate Guide to the Elements*. HarperCollins, 1996.

Carson, S. L. *Human Energy Under Varying Weather Conditions*. University of Washington, 1947.

Chinery, Michael. *Complete Guide to British Insects*. Collins, 2005.

Claridge, J. *The Country Calendar or The Shepherd of Banbury's Rules*. Sylvan Press, 1946.

Constable, Anthony and William Facey. *The Principles of Arab Navigation*. Arabian Publishing, 2013.

Cosgrove, Brian. *Pilot's Weather*. Airlife, 1999.

Dreisig, H. "Daily Activity, Thermoregulation and Water Loss in the Tiger Beetle, *Cicindela Hybrida.*" *Oecologia* (Berl.) 44 (1980): pp. 376–89.

Dunlop, Storm. *Meteorology Manual.* Haynes, 2014.

Dunwoody, H. *Weather Proverbs.* United States of America: War Department, 1883.

Eden, Philip. *Weatherwise.* Macmillan, 1995.

Elliott, S. *Nature Studies.* Blackie and Son, 1903.

Feinberg, Richard. *Polynesian Seafaring and Navigation.* Kent State University, 1988.

Geiger, R. *Das Klima der bodennahen Luftschicht.* Vieweg, Braunschweig, 1961.

Gibbings, Robert. *Lovely Is the Lee.* Dent & Sons, 1947.

Gladwin, Thomas. *East is a Big Bird.* Harvard University, 1970.

Goetzfridt, Nicholas. *Indigenous Navigation and Voyaging in the Pacific.* Greenwood Press, 1992.

Gooley, Tristan. *The Lost Art of Reading Nature's Signs.* The Experiment, 2015.

———. *How to Read Water.* The Experiment, 2016.

———. *The Nature Instinct.* The Experiment, 2018.

———. *The Natural Navigator—10th Anniversary Edition.* The Experiment, 2020.

Grace, J. *Plant Response to Wind.* Academic Press, 1977.

Greenstone, M. H. "Meteorological Determinants of Spider Ballooning: the Roles of Thermals vs. the Vertical Windspeed Gradient in Becoming Airborne." *Oceologica* 84 (1990): pp. 164–68.

Haggard, William. *Weather in the Courtroom.* American Meteorological Society, 2016.

Hamblyn, Richard. *Extraordinary Clouds.* David & Charles, 2009.

Harris, Alexandra. *Weatherland.* Thames & Hudson, 2016.

Harrison, Melissa. *Rain: Four Walks in English Weather.* Faber & Faber, 2017.

Hodgkinson, W. P. *The Eloquent Silence.* Hodder & Stoughton, 1947.

Holmes, Richard. *Falling Upwards.* William Collins, 2013.

Hunt, Nick. *Where the Wild Winds Are.* Nicholas Brealey, 2017.

Inoue, E. "Studies of the Phenomena of Waving Plants ("Honami") Caused by Wind. Part 3. Turbulent Diffusion over the Waving Plants." *J. Agric. Met.* (Tokyo) 11 (1956): pp. 147–51.

Inwards, R. *Weather Lore.* Senate, 1994.

Jankovic, Vladimir. *Reading the Skies.* University of Chicago Press, 2000.

Jefferies, Richard. *Field and Hedgerow.* Lutterworth Press, 1948.

Jones, H. *Plants and Microclimate.* Hamlyn, 2014.

Kästner, Martina, Richard Meyer, and Peter Wendling. "Influence of Weather Conditions on the Distribution of Persistent Contrails." *Institut für Physik der Atmosphäre,* Report No. 109, 1998.

King, Simon and Clare Nasir. *What Does Rain Smell Like?* 535, 2019.

Kraus, G. *Boden und Klima auf kleinstem Raum.* Fischer, Jena, 1911.

Le Blanc, J. *Man in the Cold.* Thomas, Springfield, Illinois, 1975.

Lester, Reginald. *The Observer's Book of Weather.* Frederick Warne & Co., 1964.

Lewis, David. *The Voyaging Stars.* Fontana, 1978.

―――. *We, the Navigators.* University of Hawaii, 1994.

Lewis-Stempel, J. *The Wood.* Doubleday, 2018.

Lohmann, Ulrike, Felix Luond, and Fabian Mahrt. *An Introduction to Clouds.* Cambridge University Press, 2016.

Luff, M. L. "Morphology and Microclimate of *Dactylis Glomerata* Tussocks." *Journal of Ecology* 53 (1965): pp. 771–83.

Lynch, Mike. *Minnesota Weatherwatch.* Voyageur Press, 2007.

Markham, S. F. *Climate and the Energy of Nations.* Oxford University Press, London, 1942.

Minnaert, M. *Light and Colour in the Open Air.* Dover, 1954.

Moore, Peter. *The Weather Experiment.* Vintage, 2016.

Morton, Jamie. *The Role of the Physical Environment in Ancient Greek Seafaring.* Brill, 2001.

Myers, Benjamin. *Under the Rock.* Elliott & Thompson, 2018.

Neel, P. L., and R. W. Harris. "Motion-Induced Inhibition of Elongation and Induction of Dormancy in Liquidambar." *Science* 173 (1971): pp. 58–59.

Nordhagen, R. "Die Vegetation und Flora des Sylenegebietes." *Kkr. Norske Vidensk. Akad. I, Mat, Naturv.* 1 (1927): pp. 1–162.

Office of the Chief of Naval Operations, Training Division. *Meteorology for Naval Aviators.* Government Printing Office, Washington, 1958.

Page, Robin. *Weather Forecasting the Country Way.* Penguin, 1977.

Palmer, B. *Body Weather.* Stackpole, Harrisburg, 1976.

Park, Sooyong. *The Great Soul of Siberia*. William Collins, 2017.

Parsons, K., G. Jones, and F. Greenaway. "Swarming Activity of Temperate Zone Microchiropteran Bats: Effects of Season, Time of Night and Weather Conditions." *Journal of Zoology* 261, no. 3 (2003): pp. 257–64.

Perkins, Oliver. *Reading the Clouds*. Adlard Coles, 2018.

Pinna, S., M.-F.Gevry, and M. Cote. "Factors Influencing Fructification Phenology of Edible Mushrooms in a Boreal Mixed Forest of Eastern Canada." *Forestry and Ecology Management* 260, no. 3 (2010): pp. 294–301.

Pretor-Pinney, Gavin. *The Cloudspotter's Guide*. Sceptre, 2006.

Rogers, R. R., and M.K. Yau. *A Short Course in Cloud Physics*. Butterworth-Heinemann, 1996.

Schaffer, Mary. *Old Indian Trails of the Canadian Rockies*. Rocky Mountain Books, 2011.

Scorer, R. S. "The Nature of Convection as Revealed by Soaring Birds and Dragonflies." *Quarterly Journal of the Meteorological Society* 80, no. 343 (1954): pp. 68–77.

Scott Elliott, G. F. *Nature Studies*. Blackie & Son, 1903.

Shepherd, Nan. *The Living Mountain*. Canongate, 2011.

Sillitoe, Paul. *A Place Against Time: Land and Environment in the Papua New Guinea Highlands*. Routledge, 1997.

Sloane, Eric. *Weather Almanac*. Voyageur Press, 2005.

Stewart, Ken. *The Glider Pilot's Manual*. Airlife, 2002.

Stoutjesdijk, Ph., and J. J. Barkman. *Microclimate, Vegetation and Fauna*. KNNV Publishing, 2014.

Stull, R. *Practical Meteorology*. University of British Columbia, 2015.

Thomas, Stephen D., *The Last Navigator*. Ballantine Books, 1987.

Tibbetts, G. R. *Arab Navigation*. The Royal Asiatic Society of Great Britain and Ireland, 1971.

Tovey, Bob and Brian. *The Last English Poachers*. Simon & Schuster, 2015.

Unwin, D. M. and Sarah Corbet. *Insects, Plants and Microclimate*. The Richmond Publishing Co., 1991.

Van der Poel, A. J., and Stoutjesdijk, Ph. "Some Microclimatological Differences between an Oak Wood and a Calluna Heath." *Meded. Landbouwhogesch. Wageningen* 59, no. 2 (1959): pp. 1–8.

Van Wingerden, W. K. R. E., and R. van Kreveld. "Vegetation Structure and Distribution Patterns of Grasshoppers." *Econieuws* 2, no. 4 (1989): pp. 5.

Volk, O. H. "Ein neuer für botanische Zwecke geeigneter Lichtmesser." *Ber, Dt. Bot. Ges.* 52 (1934): pp. 195–202.

Vollrath, F., M. Downes, and S. Krackow. "Design Variability in Web Geometry of an Orb-Weaving Spider." *Physiol. Behav.* 62 no. 4 (1997): pp. 735–43.

Watson, Lyall. *Heaven's Breath.* Hodder and Stoughton, 1984.

Watts, Alan. *Instant Weather Forecasting.* Adlard Coles, 1968.

Weiss, S. B., D. D. Murphy, and R. R. White. "Sun, Slope and Butterflies." *Ecology* 69 (1988): pp. 1486–96.

Werger, M. J. A. *The Study of Vegetation.* Junk, 1979.

White, Gilbert. *The Natural History of Selborne.* Penguin, 1987.

Wikström, Linnea, Per Milberg, and Karl-Olof Bergman. "Monitoring of Butterflies in Semi-Natural Grasslands: Diurnal Variation and Weather Effects." *Journal of Insect Conservation* 13 (2019): pp. 203–11.

Wood, James G. *Theophrastus of Eresus on Winds and Weather Signs.* Edward Stanford, 1894.

Woudenberg, J. P. M. "Nachtvorst in Nederland." *K.N.M.I Wetensch. Rapp.*, 68, no. 1, de Bilt, 1969.

Young, A., A. Pilar, A. Janelle, and U. Hiromi. "Wind Speed Affects Pollination Success in Blackberries." *Sociobiology* 65, no. 2 (2018): p. 225.

Acknowledgments

Early in this book I wrote that weather is a soup of air, heat, and water. A book is not a soup; it's a guided tour of an idiosyncratic fourteen-floor brick building, where every brick is different and there's a tasting platter of sushi in each room. It takes a lot more than one builder-chef-guide to get it off the ground. However, I am solely responsible for that appalling metaphor and any errors or omissions in this book.

I would like to thank my agent, Sophie Hicks, and Rupert Lancaster at Sceptre and Nicholas Cizek at The Experiment, without whom this book would not exist. This is the first of my books that was commissioned on both sides of the Atlantic at the same time. There was the potential for that to make this book harder to write, but I'm very grateful to all concerned, not least Rupert, Nick, and Sophie, that I found the opposite was true.

You may have spotted that the book sets out to do two major things: provide a comprehensive guide to the weather signs that we can sense and introduce the reader to the overlooked world of microclimates. Without Rupert and Nick's help I would not have worked out how to do that within one book and may have had to retreat from one of them.

Thank you also to Matthew Lore, Jennifer Hergenroeder, Cameron Myers, Rebecca Mundy, Caitriona Horne, Myrto Kalavrezou, Dominic Gribben, and all at Sceptre and The Experiment for your hard work, energy, ideas, and vision.

Thank you Hazel Orme for all your help in the home straight. I'm very grateful to Sarah Williams, Morag O' Brien, and William Clark for brilliant work over many years, thank you. A special thanks goes to Simon Lee, a research meteorologist, whose time and input was very helpful indeed. Many others have helped in ways that are important but not always visible: Peter Gibbs, John Rhyder, Hannah Thompson, John Pahl, Nick Hunt, Nik Huggins, thank you all.

Neil Gower's brilliant cover and illustrations are such an important part of the book. Thank you, Neil.

A big thank you to everyone who has read my books, signed up to the online course or supported my work in other ways during COVID.

I'd like to thank my sister, Siobhan Machin, for her support and valuable input.

All the names in the bibliography and sources pages helped me with ideas, facts, and inspiration, but a few deserve another mention here. Ph. Stoutjesdijk and J. J. Barkman earn my thanks and respect for their own impressive research but also their excellent work collating a wide range of academic research on microclimates. David Lewis's work in the Pacific is a lifelong help and inspiration, likewise G. R. Tibbetts on Arab navigation. Jamie Morton's work was a boon when it came to identifying ancient sources. Philip Eden, Roland Stull, and Storm Dunlop's works have all helped in slightly different ways, as have many much older ones, including those of Gibert White, Richard Jefferies, and Thomas Hardy.

Thank you also to the Camping Crew and all at the Speakeasy for bringing many smiles and laughs to 2020, online and off. I'd mention all of your names here, but then you'd have to kill me.

I'd like to thank my family for their support, especially my wife, Sophie, and sons, Ben and Vinnie, who shared COVID lockdowns with me as I wrote this book. I apologize for any wide-eyed rants about wind, dew, or the joys of being an author. It is early January 2021 as I write this final paragraph of the book, we're back in national lockdown and schools have reverted to remote teaching. This morning, after finishing our breakfast porridge, I asked my sons why there was frost on the lawn but none under the trees. This forced observation brought them no pleasure whatsoever. "Daaad! . . . . We'll be late for online registration!"

Index

Page numbers in *italic* refer to photographs and illustrations.

Page numbers followed by *n* refer to a footnote.

Greece 91, 189, 203
Grimble, Arthur 326
ground winds 57, 68
grouse 131
Gulf Stream 138
gulls 291, 331

haar 138
Haggard, William H. 197–8
hail 121–4, 155
Hales, Dr. 85
Halnaker Windmill, Sussex 108, 280
halos 40, 51, 76, 327
Hantavirus Pulmonary Syndrome 330
hardiness, plants 237–8
Hardy, Thomas 204–5, 224–3, 286
hares 131
hart's tongue fern 114–15
harvest bugs 298
Havaiki 274
Hawaii 203
hawthorn trees 211, 218, 239
Hayling Island 103
haze 140
Heart Eddies 220, *221*
heart rate, wind and 331
heat
    and cloud formation 24–5, 39
    conduction 16–17, 79
    convection 17–20, 70
    and fog 137
    heat islands 263–4, 265
    heat waves 300
    hoar frost 83
    latent heat 23–5
    ocean currents 46
    sun pockets 13–15, 15
    thermals 35–9, 69
    and wind direction 61
    see also temperatures
"heat lightning" 314
heat waves 9–10, 143
heath rush 240
heather 8, 236, 238
heathland 8, 16, 86
Hebrides 331

hedges
    shelter belts 220
    snowdrifts 130, 172
    and winds 73, *73*, 290, *291*
heiligenschein 76
helm wind 102
herd animals 287
herons 80
high-pressure systems 44
    blocking highs 9–10, 12
    crosswinds rule 205
    fog 135
    hoar frost 83
    local winds 203
    sea breezes 198
    and snow 127
    wind direction and 61–2
hills
    and cloud-base colors 279
    foehn winds 102–3
    and frost 86–8, *88*
    and rain 99, 100–2
    split winds 189–90, *191*
    summit winds 197
    see also mountains
Himalayan Balsam 237
hoar frost 81–2, 83–90, 307
Hodgkinson, W. P. 82
holly oak 210
holm oak 210
Homer 189, 205
honeyeater bird 289
hoodoos 250, *253*, 254
horses 207, 285, 286–7
House of Wild Truth 111–13
houses, and winds 193
humans, scent 115–16, 117, 120, 251
humidity
    air masses 44
    and clouds 35
    "humid blanket" 25
    lichens and 245–6
    and scents 233–4
    signs of 268–9
    and snowflakes 125
    summer storms 306
    winds and 59–60
    woodland hygrometers 215–16

landspouts 318
Lanzarote 185
Lapland 130
larch trees 211, 212
latent heat 23–5
Latin America 246–7
leaves
    animal tracks 111
    and drought 215
    microclimates 240–2, *242*
    and rain 92–3, 213
    red colors 226–7
    shapes 242
    size of 241
    sounds of trees 223
    texture 241–2
    and wind 59, 221–2
Lee, Simon 289
lens clouds 161–3, *162*, 195, 276
levant wind 64, 183
Levanter wind 183
Libya 102
lichens 80, 110, 131, 215, 220, 225,
    244–6
light
    brightness 77–8
    butterflies and 296
    from stars 324
    and frost 90
    fungi and 243
    halos 76
    red colors 226–8
lightning 123, 178, 301, 313–17
Liu Chi 223
local winds 182–96, *186*, 191, *194*,
    201–6
lofting smoke *119*, 120
London 266, 305
looping smoke 118, *119*, 120
Loughborough 144
Love Lakes, Dubai 172
low-pressure systems 47, *48*, *49*
    cirrus clouds and 148, 152–3
    crosswinds rule 205–6
    hurricanes 318
    jet stream and 151
    wind direction and 61–2, 67

lunar calendar 327

mackerel skies 158–60
Maine 236
Maldives 274–5
Malus sieversii 227
mamma, storm clouds 309, 317
Manhattan 259–61, 267, 270–1
mares' tails 148, 155
marigolds 234
Massenerhebung effect 217
matgrass 235, 235
Matterhorn 165
Maya 332
Mediterranean 64, 183, 187
Melanesia 273
Mendip Hills 100
metal, heat conduction 16–17
meteorological instruments 89
meteors 324–5
Mexico 229
Miami 104, 231
micro-seasons 239
microclimates 7–9
    coastal 282
    leaves 240–2, *242*
    snow 130
    trees 208, 227
    woodlands 221–2
Micronesia 27, 273
microplastics, in rain 91
Middle East 266
midges 293, 297
Milky Way 325
mist 26, 134, 322–3
    see also fog
mistral 187
Molière 204
Monroe, Marilyn 261
Monroe Effect 261, 270
monsoon 62, 275–6
Montana 103
moon
    full moon 77, 173
    halos 40, 50, 327
    weather lore 322–3
moschatel 239–41

About the Author

TRISTAN GOOLEY is the *New York Times*–bestselling author of *How to Read Water, How to Read Nature, The Natural Navigator, The Lost Art of Reading Nature's Signs,* and *The Nature Instinct.* He is a leading expert on natural navigation, and his passion for the subject stems from his hands-on experience. He has led expeditions in five continents; climbed mountains in Europe, Africa, and Asia; sailed boats across oceans; and piloted small aircrafts to Africa and the Arctic. He is the only living person to have both flown solo and sailed single-handedly across the Atlantic, and he is a Fellow of the Royal Institute of Navigation and the Royal Geographical Society. To see more from Tristan Gooley, please visit his website, naturalnavigator.com, and follow him on social media.

NaturalNav | @thenaturalnavigator | thenaturalnavigator

*Watch for storms
when clouds are more
tall than wide.*

*Gliding birds mean
stable air and thus,
fair weather.*

*Blackberries grow more segments
where the wind is faster.*